The Great Driving Right Show

About Salvage Editions

Salvage Editions is a collaboration between Verso Books and *Salvage*. Edited by the Salvage Collective, the series publishes writers the editors admire on various topics. Some of these interventions extend and develop arguments initially published in Salvage, others are entirely new. As with all *Salvage*'s publishing, Salvage Editions intervenes in the key theoretical and political questions thrown up by our moment in ways both politically incisive and stylistically ambitious and engaged. The Salvage Collective does not believe, put simply, that radical writing should not also strive for beauty.

Since 2015, *Salvage* has been publishing essays, poetry, fiction and visual art in its print edition, currently published twice a year. In 2020, along with the Salvage Editions book series, the Salvage Collective also began a live online events series in collaboration with Haymarket Books, called Salvage Live.

Over the years and issues, a cluster of concerns has emerged as core to *Salvage*'s project, including global political economy; modern political subjectivity; the social industries; sexuality, race and identity; and eco-socialism.

For more on *Salvage* and for information about subscribing, please visit the website at salvage.zone.

The Great Driving Right Show

Cars, Crisis, and the Rise of Fossil Fascism

The Zetkin Collective

VERSO
London • New York

First published by Verso 2026

The manufacturer's authorized representative in the EU for product safety (GPSR) is
LOGOS EUROPE, 9 rue Nicolas Poussin, 17000, La Rochelle, France
contact@logoseurope.eu

1 3 5 7 9 10 8 6 4 2

Verso
UK: 6 Meard Street, London W1F 0EG
US: 207 East 32nd Street, New York, NY 10016
versobooks.com

Verso is the imprint of New Left Books

ISBN-13: 978-1-83674-097-1
ISBN-13: 978-1-83674-099-5 (US EBK)
ISBN-13: 978-1-83674-098-8 (UK EBK)

British Library Cataloguing in Publication Data
A catalogue record for this book is available from the British Library

Library of Congress Cataloging-in-Publication Data
A catalog record for this book is available from the Library of Congress

Typeset in Fournier by MJ & N Gavan, Truro, Cornwall
Printed and bound by CPI Group (UK) Ltd, Croydon, CR0 4YY

Contents

Introduction

Apocalypse is in the air. Over the summer of 2023, a group of 'Blade Runners' started sabotaging the infrastructure of ultra-low-emission zones (ULEZs) in and around London. Clad in black, face to toe, they targeted the cameras that scan number plates on cars determining whether they comply with the minimum environmental standards or must pay the daily charge of £12.50. Some Blade Runners dismantled the structures entirely, removing the cameras from their perch and stashing them away in lock-ups. Others hacked the posts down, allowing gravity to smash the contraptions on the pavement. The most deft, however, arrived on the scene at night with balaclavas and long-handled secateurs to snip the cables before making a ghoulish exit. One Blade Runner recorded himself in the act, addressing the powers he was resisting: 'It might take your blokes half a day to put one up. It takes me less than a minute to take one down. So fuck your fucking ULEZ, shit cunts. This is our country and we're taking it back.'

What exactly was motivating the Blade Runners' sabotage? Browse the online spaces where the most committed congregate and you'll see it has nothing to do with air quality and little to do with the charges involved. These are spaces where all manner of contemporary social anxieties and moral panics collapse into one another. Railing against vaccines, 'climate lockdowns', 'gender ideology', 'the deep state' and digital IDs, the Blade Runners see themselves as freedom fighters, resisting a totalitarian state with a global agenda. Yet it would be a mistake to dismiss these actors as cranks on the margins of society. While they partly target developments of traditional concern for the left – such as the expansion of state and corporate surveillance technologies or the privatisation of responsibility through fees and fines – their diagnosis and prescriptions drive politics into new terrains. At once transforming and mainstreaming far-right discourses, they synthesise fossil nationalism with a radical form of libertarianism. Although this represents but one among countless cases of a pandemic-era politics that continues to mutate today, the Blade Runners' strategic contestation of decarbonisation policies reflects a new frontier of reaction and a new phase of denial.

As the far right overwhelms the news cycle, crisis after crisis, it is easy to forget that before 2020 a very different kind of global mobilisation was on the rise – one animated by the conviction that a liveable planet must be defended, not disavowed in the name of 'freedom'. In the early 2010s, the COP summits became international sites of mass mobilisation, while waves of indigenous and environmentalist resistance targeted local sites of extraction: mines, pipelines, ports and forests. But it was the late 2010s that saw the climate movement grow into a truly global force. Greta Thunberg began her solitary school strike in 2018, inspiring a wave of international protests including the massive Fridays for Future student movement. By occupying Speaker Nancy Pelosi's congressional office, among other actions, the US-based Sunrise Movement broadcast the Green New Deal to national audiences and beyond. Coming out of the UK, Extinction Rebellion (XR) developed forms of civil disobedience aimed at climate consciousness raising. In

2019, millions participated in globally coordinated climate strikes, with significant demonstrations in Angola, Chile, Pakistan, and the Philippines. The global surge culminated in September with climate strikes counting around 7 million people across 150 countries – 300,000 in the UK, half a million in the US, 1 million in Canada, and 1.4 million in Germany – demanding action ahead of the UN Climate Summit.

The COVID-19 pandemic choked off the momentum. Climate justice mobilisations contracted into a digital echo: demonstrations transformed into webinars, mass actions into strategy calls, Fridays for Future into Fridays on Instagram. At the same time, the pandemic produced fertile ground for very different kinds of anti-establishment mobilisation, which swarmed onto Telegram channels, alternative media and the streets. Fuelled by uncertainty, economic strain and distrust in institutions, a right-leaning parallel political universe increasingly repurposed the language and ideological frames of resistance to fit an authoritarian–libertarian mould. Meanwhile, international centre-right and far-right parties not only learned to absorb the populist surplus of such movements; they also took cues from progressive opponents and went on the offensive themselves, conjuring crises of apocalyptic proportions. What followed was an entire series of moral panics that, taken together, constitute a new phase of far-right denial and an astonishing era of 'greenlash'.

In the framing of far-right moral crusaders, the world appears divided around a set of binaries: truth and lies, nationalists and globalists, freedom fighters and totalitarians, producers and parasites, the suburbs and the city, workaday motorists and climate protesters. Fuelling the far right's conspiracy engine, such binary images and narratives construct what we call an 'inverted crisis'. This inverted crisis at once mirrors the framing used in progressive struggles and obscures the causes and consequences of planetary warming. The crisis is not climate change, it is said, but what 'they' – the 'globalist' elite and 'woke' mobs – are going to do about it. For far-right movements and parties, proposed solutions to the ecological crisis are the crisis itself.

Most accounts of fascism suggest that some kind of crisis is required as its condition of possibility. As they deepen, interlocking ecological, economic and social crises – or what we call the 'organic crisis' of contemporary capitalism – are ripe for the emergence of fascist politics. The inverted crisis is the far right's conjunctural response to an accelerating organic crisis. Through the radicalisation of the far right, crisis itself becomes a mechanism of hegemonic disruption. *The Great Driving Right Show* examines the spectacular success of the far right in constructing conspiratorial crises and combating the progressive push towards a green transition in the post-pandemic conjuncture. Drawing from a number of international cases, we show how an anti-ecological and pro-car politics unites diverse actors, mainstreams once fringe positions and drives the entire political spectrum to the right. As the far right undergoes a process of fossil fascisation, our climate and politics are taken to new extremes.

Mutations of denial

Climate denialism – the organised obfuscation of anthropogenic climate change – has become a central category in the lexicon of critical climate politics. The ABC of climate denial has gone something like this: 'if there is no warming trend; or if it cannot be attributed to humans, or if the impacts are harmless or even beneficial, then there is no need for any action'.[1] The concept of denial has been applied in studies to understand the formation and effects of historical propaganda campaigns, to account for the agents and strategies of misinformation, to explain why public appreciation of the problem of climate change lags scientific consensus, and to identify the cohesive force behind the 'climate countermovement' – that is, the constellation of institutions and actors operating on behalf of vested interests to systematically mislead the public and influence the political process.

Since climate scientists sounded the alarm in 1988, the forms and strategies of denial have mutated time and again. A periodisation of climate denial can be found in *White Skin, Black Fuel*, which tracks different phases of its development vis-à-vis the shifting political headwinds of the time: the 'early corporate' phase (1989–2001), the 'late corporate' phase (2001–7), the 'conservative' phase (2007–16) and the 'far-right' phase (beginning with the first Trump presidency), with more lines of continuity than breaks between them.[2] The early corporate phase was a caricature of vulgar Marxism, in which ideology served as a direct expression of economic interests. Fossil fuel companies, whose line of business was imperilled by the findings of climate science (including their own in-house models), transparently disseminated a swath of materials that directly rejected those findings: the climate is not changing, carbon dioxide is good for you, and so forth. This strategy – so clearly motivated by self-interest and running counter to a growing public consensus – proved unsustainable. The late corporate and conservative phase was therefore marked by the veiling of these underlying motives: 'dark money' was channelled through donor funds and think tanks who could preach the fossil fuel gospel instead.

The arrival of the far-right phase of denial marked a rupture. Parties and actors with less direct interests in the fossil economy repeated ideas and tropes first seeded by fossil fuel companies in the late 1980s. In other words, climate denialism detached from the base and established itself among the general ideologies of the far right. This mutation has proved remarkably successful. Through the rise of far-right parties, climate denial has made deep inroads into European politics for the first time and is now heard in the parliaments and chambers of countries previously considered strongholds of climate rationality. Seemingly, the greater the distance from the economic base – and any obvious pecuniary motivations on behalf of the oil and gas industry – the greater potential for denial to gain political traction.

In this conjuncture, we can find five novel and interrelated processes that point towards the emergence of a fifth phase of denial, or what we might call 'the late far-right phase', starting with the pandemic and accelerating in its wake. Each of these five intertwined processes is not reducible to denial, but is crucial to understanding how it works today. All five constitute the hallmarks of our conjuncture, the late far-right phase.

First and most visibly, the far right advances through a double movement of radicalisation and mainstreaming. In popular discourse and government policy, many things dismissed or deemed unfathomable only a few years ago are now commonplace, with anti-migrant and anti-climate agendas increasingly widespread and durable. As the far right is normalised, it is radicalising rather than moderating. Concepts like 'remigration' – which demands the deportation of millions of people, spanning from 'illegal migrants' to 'unassimilated' citizens – have moved from the extremist fringes to the official platforms of parties edging closer to power. This runs from the election of Giorgia Meloni in Italy to the Alternative for Germany (AfD), whose 2025 rise atop the polls represents a first for an extreme right party since the formal end of fascism in 1945. International laws meant to protect asylum are being widely diminished or ignored, while jails in Central America and Africa fill up with those expelled from the US who have committed no crime, many of them not even citizens of the countries to which they have been deported. If 'the wall' was the icon of the fourth phase of denial, the masked ICE agent, the unmarked van and the 'return ticket' of proposed mass deportations are the symbols of the fifth.

Second, this radicalisation takes an epistemic form, with conspiratorial logics seeping into and reshaping the establishment right, all while traditional conservative parties weaken and the political centre implodes. In this process, the far right rises through a moral panic cycle that exploits existing social anxieties and turns an ever-expanding cast of 'deviant' subjects into scapegoats. Over time, as individual moral panics build on one another, they converge into a 'general conspiracy' imagined as the arrival of 'Armageddon'.[3] We call this *the inverted crisis* of the far right. In this framing, it is not climate change but climate mitigation that represents an existential threat to the rightful inhabitants of the West. Through strategies of inversion, forms of denial directed at the climate crisis are stitched together with 'crises' of migration, race, gender, the family and the nation. Such articulations often present themselves as a topsy-turvy mirror image – as an inversion of liberal framings and leftist critiques. From this vantage, 'green new scams' are set to weaken the industrial fortitude of the nation, climate activists are the real agents of disaster and innocuous (and inadequate) green traffic schemes are harbingers of a globalist plot to secure a totalitarian world government. Amid the ongoing hegemonic breakdown, the far right thus turns from defensive to offensive strategies of denial and delay. The mass proliferation and uptake of conspiracy theories at the height of the pandemic outpaced the dreams of even the most hopeful far-right actors in years prior.

Third, these processes of radicalisation, conspiratorial absorption and mainstreaming find increasingly overt expression in the streets. Whereas, in previous phases, denial was largely situated in boardrooms, think tanks, party manifestos, and mainstream and social media, far-right actors now seek to mobilise protest movements with anti-climate impulses. Often pointing to the *gilets jaunes* for inspiration, and spreading rapidly via social media, corporate and political actors compel the public to move onto the streets. While coordinated, these movements retain a politically indeterminate character. Partially spurred on by 'corporate grassroots' tactics, they go beyond mere 'astroturfing', steered instead by a relatively autonomous – and primarily petty bourgeois – base. These are the social forces through which the mediatised moral panic cycle touches down in protest cycles. Some aim to obstruct sustainable-traffic schemes, as in the anti-ULEZ and anti-fifteen-minute-city movements across Britain, or to embed anti-ecological demands under the aegis of the 'freedom' movement against COVID-19 restrictions. Others still demand the reinstatement of fertiliser and fuel subsidies, as with the farmers driving tractors across the capitals of dozens of European cities. And, of course, some even contest the results of elections, as in Washington and Brasília. In all cases, protest has become a powerful feature of the overlapping denialist and far-right repertoires.

Fourth is the backlash to and subordination of climate, once ascendant as a political concern in the 2010s, in the hierarchy of crises. The stage is crowded with calamity: the pandemic, inflation and the cost-of-living crisis; the outbreak of war in Europe and the Middle East; Israel's genocide in Palestine; the build-out of AI, and finally the crisis of liberal democracy itself, with the far right increasingly capable of achieving significant electoral gains and setting the tone of political discourse. In this context, climate is increasingly pushed to the margins of the political agenda, movement energy is rerouted, and talk of transition is blunted. Part of this subordination is driven by reactive psychopolitical affects – a backlash to the earlier momentum of climate movements and policies that threatened fossil-fuelled ways of life. These various crises, in turn, are expressed through the punitive targeting of the left, further hobbling the already diminished resources on which the climate movement relies.

Fifth and finally, fossil capital, facing minimal political resistance and gaining the legitimacy granted by heightened security concerns and economic anxiety, is reproducing itself with renewed ease, freely funnelling super-profits into new fossil investment with restored voracity and locking in denial at the level of political economy. Financial institutions are playing their role, with the world's biggest banks committing nearly a trillion dollars to the fossil fuel sector in 2024 alone. In previous phases of denial, the representatives of oil and gas were broadly compelled to conceal their operations beneath a green veneer. With fears of stranded assets now diminished, such façades can be dropped, ties to renewable investments severed, climate alliances abandoned and fossil operations expanded with limited restraint. With overshoot for today and geoengineering for tomorrow, the fantasy that market innovation or technological inventions will fix the problem holds sway.

In the following pages we examine these hallmarks, showing how specific forms and strategies of denial have adapted, mutated and accelerated. Our theoretical approach to denial is broad, referring not simply to the outright rejection of climate science or to a moralistic or individualistic matter of belief. As the challenge of decarbonisation is political and economic in nature, it is imperative to understand the structural and organisational forms that denial takes. Increasingly, strategies of denial intertwine with conflicts over identity, race, gender and class, demanding various tools of analysis. We thus use denial as a lens to analyse the discursive, affective and psychoanalytic features of social formations that reinforce the drivers of climate breakdown.

Key to the far right's rise in this conjuncture are the politics of automobility and the internal combustion engine. Organised climate denial has, of course, been deeply enmeshed with automobility from the beginning, with auto companies Chrysler, GM and Ford joining oil companies as the main financiers of the earliest denialist organisations. If climate science offers an 'immanent critique of the industrial base of modernity', and if climate denial has functioned as the defence of that base, then cars, as one of the most ubiquitous commodities of industrial modernity, are sure to be a key site in any denialist politics.[4]

In and beyond the pandemic years, however, fossil-fuelled automobility encountered its most serious threat yet, triggered by increased adoption of 'green' urbanist policies, climate regulation and formal targets for combustion engine phase-outs. Thus, central to the late far-right phase is the so-called 'war on cars', a discursive framing that represents any initiative to reduce car dependency as a malicious attack on individual freedom.[5] Through this framing, the far right cathects popular energies to the internal combustion engine, from the diesel engines of farmers to the daily commute of the self-reliant suburbanite. Confronted by environmental regulations, sustainable-traffic schemes, and rising fuel prices, the discourse of the embattled motorist falls within a pattern of resistance used by car and oil companies to mobilise drivers against corporate regulation and taxation.[6] It is also central to contemporary carbon populism, which represents ordinary, fossil fuel-consuming citizens and extractive industries as existentially threatened by globalist elites and the radical 'woke' left, and therefore in urgent need of defence.[7] Through the 'war on cars', the far right finds another way to radicalise its rhetoric and mainstream its politics, turning cultural grievance into electoral capital. In a populist key, the car and the driver are synonymous with 'the people'.

Driving to the top

Taking a broad sweep of the five years since the pandemic suggests a period of waning and then waxing for the far right. Initially, the epochal rupture of the pandemic pushed the far right onto the back foot, signalling a plateau in their fortunes, which, since the middle of that decade, had been riding high. With the likes of Trump, Bolsonaro and Johnson bumbling and bungling their way through the pandemic – leaving inflated death tolls in their wake – centrist managerialism momentarily had an advantage, viewed as the steadier pair of hands to guide society through such a crisis. Biden's election victory in November 2020 seemed to only confirm the sea change. But as interactions moved online, conspiracism became the lingua franca. The far right tapped into these online subcultures, speaking the language of freedom – of bodily autonomy and individual rights – and adapting progressive idioms for reactionary ends.

As lockdowns turned into second lockdowns, and vaccines into vaccine passports, the authority and legitimacy of governments, science and journalism were under sustained attack. Institutional distrust became a dominant structure of feeling. Compounding this mood, energy, inflation and cost-of-living crises hit Europe, also bringing distributional conflicts to the fore. With pressures from climate displacement, labour shortages, and wars in Europe, the Middle East and Africa – as well as far-right agitation – the border returned as a primary site of political contestation, following its temporary retreat to the background during the pandemic peak. Amid the upheaval, the far right could occasionally be heard championing a version of material politics, with some cosplaying as the voice of the working class. While refugees and migrants have long been blamed for falling wages, high unemployment and shrinking budgets, climate activists and their elite agenda increasingly became the culprits for vanishing jobs, economic precarity and endangered energy security.

This mounting distrust of establishment actors crystallised in a political crisis of incumbency: a wave of electoral collapses for sitting governments around the world, seemingly prefigured by the chainsaw-wielding Argentine Javier Milei in late 2023. In Europe, Macron, Scholz and Sunak were among the highest-profile dethronements. All told, incumbent parties lost 85 per cent of races in the 2024 super-year of elections, a figure that stood at just 25 per cent two decades ago. In the EU elections, the far right surged while liberal and green parties waned, with Rassemblement National (RN) receiving more than 30 per cent of the vote in France, before going on to win the first round in national elections later that month. Its ambitions were frustrated only by the emergence of a 'popular front': an anti-fascist multiparty electoral strategy, coordinated by the left, to block the far right from power.

The re-election of Donald Trump consolidated the resurgence of far-right nationalism. From 2025 into 2026, the administration bolstered the power of Homeland Security, boosted ICE's funding by $170 billion and brought brutal raids to the homes of migrants and the screens of millions in a growing spectacle of cruelty. Following the assassination of Charlie Kirk, the establishment media memorialised the far-right activist, with tens of thousands rallying around their 'martyr' at a stadium-filling memorial featuring full-throated Christian nationalism, and the administration began an aggressive phase of political demonisation and media censorship. After the July 2025 'No Kings' protests brought around 7 million to the streets, Trump officially declared 'antifa' a domestic terrorist organisation while semantically expanding 'antifa' to mean all dissenting liberals and leftists with 'anti-capitalist' or 'anti-Christian' views. Beyond its borders, the administration turned its attack towards leftist organisations in Germany, Italy and Greece: the 'anarchists, Marxists, and violent extremists of antifa' purportedly waging campaigns of terror against the US and its allies.

Amid the heightened repression and paranoia, international far-right alliances strengthened. Alongside Big Tech billionaires, also on display at Trump's inauguration were far-right leaders like Javier Milei, Giorgia Meloni (Brothers of Italy), Nigel Farage (Reform UK), Santiago Abascal (Vox), Tino Chrupalla (AfD), Éric Zemmour (Reconquête) and Filip Turek, the Nazi-nostalgic Czech leader of the Motorists for Themselves party. In keeping with the spectacular register of the moment, glitzy conventions, CPAC conferences and 'patriot summits' became key organisational nodes for the global far right. While JD Vance promoted AI investment alongside far-right politicians in Europe, Musk financed Tommy Robinson's legal battles and beamed into rallies in Germany and Britain, where over a hundred thousand marched at an anti-immigration demonstration. Riding the wave of this internationalisation, concepts like 'remigration' now circulate freely between Europe and the United States, each context reinforcing and radicalising the other.

Seemingly buoyed by the magnetic force of a Trump White House, the far right has risen to the top of the opinion polls in the largest economies in Europe: Germany, France and Britain. In Germany, the 'firewall' around the far right has continued to crumble. Under Alice Weidel, the AfD leads the polls, marginally ahead of the CDU, while Chancellor Friedrich Merz has repeatedly blurred the lines between the two parties by leaning into much of the AfD's rhetoric. Seemingly undeterred by the revelations of Nazi affiliation and affection within its ranks, the AfD actively pushes some of its more extreme figures to the forefront. Weidel, for example, finally came to see Bjorn Höcke – a charismatic extremist with a deep neo-Nazi past and active links to the country's far-right youth movement – as a figure to celebrate and a suitable candidate for a ministerial post.

In Britain, Reform UK's rise has been meteoric. Revoking the indefinite right to remain and pledging the expulsion of hundreds of thousands of individuals have become policy pillars that even the Tories have embraced. Farage's success has been mirrored in the collapse of support for both Labour and the Conservatives. Conceding to the opposition, both establishment parties have made pivots to the right, particularly in their hardening stances on migration and the policing of protest. While more 'intellectualised forms of ethnonationalism' may circulate online, it appears that Farage does not necessarily need them. Rather, what is fuelling the success of Reform is a 'folk wisdom that gestures loosely towards fraud, asylum, government waste and welfare cuts'.[8] The far-right international now shares an anti-statist language popularised by Milei, Musk and DOGE.[9]

As temperatures rise, politics is dancing to a new tune: the far right increasingly decides the problem, sets the rhythm and calls the steps. Attempts to curtail far-right momentum by the liberal centre – through legal challenges, surveillance, or cordons sanitaires – now give way to policy imitation. Over the same period, the far right has established a foothold among younger voters, particularly young men, while effectively absorbing online language and subcultures. Leaders like Farage and Bardella, for instance, have millions of followers on TikTok. In Germany, young people are now twice as likely to vote for the AfD as they were at the start of the pandemic. By the middle of the decade, far-right parties and leaders no longer operate as illegitimate outsiders but rather as the gravitational centre of politics.

Greenlash

For a brief window, it appeared as if the pandemic might be a boon for climate action, with sizable portions of recovery packages earmarked for climate-related initiatives. As the economy slowed under lockdown, so too did emissions. But when restrictions were lifted, so was the green veil. Emissions quickly returned to pre-pandemic levels, and Russia's subsequent invasion of Ukraine ignited an already unstable global energy system, sending the price of oil and gas rocketing from historic lows to magnificent highs. An avalanche of new investments in fossil fuel infrastructure was quickly set in motion. Rapidly, the imperatives of national security and the rising cost of living pushed climate ever further down the descending hierarchy of crises as incumbent governments were ousted around the world, none more significantly than the administration of Joe Biden.

On day one, Trump signed a slew of executive orders. With a stroke of the pen, the state withdrew from the Paris Climate Accords, halted offshore wind projects, denied the rights and very existence of trans people, barred the practices of racial equity or 'DEI' and abolished all regulations on energy-guzzling AI innovation. In the view of the new Energy Secretary under Trump 2.0, Chris Wright, Biden's 'quasi-religious' climate policies would be terminated to make space for the 'unabashed' pursuit of 'more American energy production and infrastructure'. Over the first year in office, the administration revoked regulations, prohibited renewable energy projects and gave handouts to the oil and coal industries. Many attacks homed in on mobility. In San Diego County, million-dollar schemes 'hostile to motor vehicles' were scrapped, while elsewhere cycle lanes were decried as 'European and un-American communist garbage'. Trump personally waged war on New York's bicycle lanes ('so bad!') and 'destructive' congestion pricing, which conspired to prevent drivers visiting the city. Announcing efforts to derail the latter, he posted a fake image of himself on *Time* magazine, wearing a crown, captioned, 'CONGESTION PRICING IS DEAD. New York is SAVED. LONG LIVE THE KING!'

Yet in this 'late far-right phase', this backlash hardly belongs to the far right alone. Centrist governments, nominally committed to capitalist climate governance, normalised the rollback of climate commitments as common sense, presenting regression as responsibility. In Britain, the Labour government mulled over major oilfields in the North Sea while also approving airport expansions at Heathrow and Gatwick, enabling 100,000 extra flights at the latter each year – what the Transport Secretary described as a 'no-brainer' for economic growth. In Canada, Liberal Prime Minister Mark Carney dismantled nearly all major climate policies from the Trudeau decade. In Brussels, representatives of the German CDU actively lobbied to stall the EU's planned phase-out of combustion engines and give car makers 'breathing space'. Meanwhile, EU conservatives abandoned their traditional alliance with centrists to vote with the far right to weaken environmental regulations on large businesses.

If backlash refers to a sudden and aggressive political reaction attempting to reverse a political development, then greenlash is the form this takes when the legitimacy of climate commitments is contested.[10] Looking at the surface, some environmental researchers suggest that the extent of any greenlash is overblown, since support for climate policies remains high across a number of political contexts. But when and where grievances erupt, we consistently find far-right parties, the foremost defenders of the fossil-fuelled privileges, proving themselves adept at amplifying and generating political momentum from the fallout. Industry also adjusts itself to the political climate of greenlash. In 2025, the oil and gas giants gleefully shed their green pretensions as BP announced that the company's push for renewables had been 'misguided' and thus restored their 'unwavering focus on growing long-term shareholder value'. Equinor, meanwhile, halved its renewable investments for the next two years, and Shell halted any further investments in wind generation.

If many political leaders are now embracing the fact that decarbonisation is 'just too hard', then ideological comfort or moral consolation might be realised by downplaying or diminishing the severity of the problem. With the climate question effectively shelved, focus is shifted to the nation – its borders, industry and 'values'. Such a scalar retreat by centrist forces – a strategy of containment analogous to the hardening of discourse on immigration – is serving as a boon for the far right. Outflanked to their right, establishment parties shift from making green appeals to embracing climate 'realism'. The far right's unflinching ranking of the nation over the climate forces the political imagination back into the bounds of the nation state, with all the predictable consequences for a crisis that demands transnational cooperation.

Driving through the storm

The compounded shocks of climatic chaos – fire, flood, hurricane, heat, pandemic – have begun to reshape not only the planet's physical landscapes, but also its political, economic and social orders. A decade after the Paris Agreement, the instabilities connecting the Earth system and the capitalist world system continue to deepen. Current national pledges place the world on course for roughly 2.6 °C temperature increase by 2100, far exceeding the Paris targets of limiting temperature rise to between 1.5 and 2°C. Such a level of warming would push the Earth into a climate outside all historical human experience – one defined by an accelerating spiral of climate violence. Yet capital continues to prioritise short-term profit and asset protection over long-term planetary stability.

Nowhere is this impasse more clearly embodied than in the car industry – the emblem of fossil modernity – where the drive to preserve business as usual remains inseparable from the drive towards catastrophe.[11] Globally, half of total oil use is accounted for by road transport. In several countries in the capitalist core, including the United States and the United Kingdom, transport is the largest source of emissions, with passenger vehicles accounting for the majority. Moreover, road vehicles – alongside aviation – often remain the only sources of emissions that continue to increase annually.[12] Only counting end-use emissions and ignoring those in manufacturing and associated infrastructure, however, drastically underestimates the ecological cost of auto dependency. The auto industry's steel consumption, for example, accounts for one-tenth of total global steel demand, and since most steel is produced with coal, nearly one per cent of global annual emissions come from just the auto industry's steel usage. The upward pressure on both direct and embedded emissions will only intensify: currently the world is populated with around one and a half billion vehicles, with 80 million extra getting added to the fleet each year.

If the Great Acceleration denoted the energy- and resource-intensive patterns of consumption and production driving the postwar boom, then during the first decades of the twenty-first century fossil fuel consumption found an extra gear, with carbon emissions around the world accelerating at pace.[13] SUVs were the second-biggest factor in global emissions increases, behind only the power sector, and ahead of shipping, aviation and heavy industry. Today, if SUVs were a country, they would rank as the fifth-highest emitter in the world, only surpassed by the United States, China, India and Russia. In the advanced capitalist core, SUVs now account for over half of new car sales, with European consumers no longer bucking the trend. The increasingly affluent middle classes in the rest of the world, too, are rapidly catching up. Many of these vehicles are comparable in size to tanks on the battlefields of the Second World War. The bloating of car size is now described by the International Energy Agency as the 'defining automobile trend of the early twenty-first century'.

The heavier cars get, the more damage they do to the road: the so-called 'fourth-power rule' describes how road wear increases exponentially with every extra unit of weight. And the heavier cars get, the more plastic particulates their tyres shed. Over the course of its useful life cycle, a tyre loses one-tenth of its mass due to abrasion, the particulate finding its way into the soil, water and air. With over 2 billion produced annually, tyres thus constitute a leading source of microplastic pollution. Due to their highly complex and toxic chemistry, the synthetic rubber wheels can emit a hundred times more volatile organic compounds than an average exhaust pipe. Unsurprisingly, the tyre industry treads the same ground of the denialist apparatus, working to manufacture scientific doubt and otherwise delay or weaken regulation.

This feedback loop of deterioration and decay – bigger cars, balding tyres, broken roads – demands ever more resources and energy to maintain: more coal to make the cement, more cement to fill the gaps, more oil to make the ethylene, more ethylene to make the tyres. Residual infrastructures lock social and ecological relations into place, conferring a certain inertia upon auto dependency and its destructive spirals. As a result, we burn from, choke on and even eat automobility's toxic by-products.

In a dialectical turn, the climate agenda, and the science of climate change, have posed the single most sustained challenge to the internal combustion engine. Long before the details of global warming crystallised in civil society, GM and Ford knew all about the consequences of unbridled carbon emissions – their research departments documenting the greenhouse effect since the 1950s. As climate research expanded and consensus around its findings consolidated in the 1990s, tougher climate regulations were imposed on manufacturers, mostly in the form of energy efficiency standards. Fast out the blocks, California's 1990 zero-emission programme forced producers to offer electric models, while Norway, another frontrunner, provided incentives to encourage consumers to switch. Loyal to a world dominated by the automobile, capitalist climate governance prioritised one solution above all others: to switch out the internal combustion engine for a lithium ion battery.

Ecological critiques of EVs are not hard to find. Most focus on the demand for rare earth and critical minerals, noting the already strained and often violent supply chains for lithium, cobalt, nickel and graphite.[14] Battery production is particularly resource-intensive, typically accounting for around half of any given vehicle's embedded emissions. In China, the foremost producer of electric vehicles, one estimate suggests that a single car demands two and half tonnes of coal to build.[15] The vehicle, moreover, is only as green as the grid that powers it. The electrification of private mobility will boost overall electricity demand, which, alongside the massive build-out of AI server infrastructure, threatens to overpower outdated national power grids. With transport the paradigmatic sector to show how efficiency gains are offset by rising demand, one must expect continued growth in transport volumes to increase overall energy consumption, in turn intensifying demand for resources upstream. Even with full electrification, the US would not meet its emissions targets, as shown by an article in *Nature*. What is needed, the authors thus conclude, is 'a wide range of policies that include measures to reduce vehicle ownership and usage'.[16] In short: to decarbonise transport, traffic must be reduced.

These material contradictions have political corollaries. Schemes to limit fossil-fuelled automobility – such as the phase-out of internal combustion engines, low-emission zones, reduced-traffic areas or the removal of fuel subsidies – have the capacity to enrage people across the political spectrum. A central tension here revolves around class: on both left and right, the defence of the status quo is often cast as the defence of the working class. Car use, however, is less often a matter of choice than of structural coercion. Many drivers depend on their vehicles because of the spatial separation of work and home and the chronic under-provision of public transport.

Limits on established norms of mobility thus appear to threaten livelihoods tied to these forms – not only those directly employed in the automotive sector, but also the wider web of car-dependent trades and services. Policies such as carbon taxes, fuel duties, congestion charges or the requirement to upgrade to a newer and more energy-efficient model can therefore seem punitive and regressive, with the burden of decarbonisation falling most heavily on those already exploited.

Equally true, however, is that the poorest households often lack cars altogether and that, in urban cores, air pollution exposure correlates with racialised and classed inequalities. When anti-car policies are coupled with serious investment in public transport, the reduction of fossil fuel mobility can in practice be transformational for parts of the working class. Nothing intrinsic fixes the class character of anti-car schemes. Zooming out beyond national borders, the fuller contours of the problem come into view. The global impacts of climate breakdown driven by automobility – heat, floods, displacement – fall disproportionately along existing uneven geographies of inequity. From a climate justice perspective, the imperative is unambiguous: an absolute reduction in numbers of private cars.

What is ecologically urgent, though, is rarely politically convenient – and these tensions form the terrain that the far right has learned to exploit.[17] If the programme for climate mitigation can, at a high level of abstraction, be boiled down to a simple maxim – 'decarbonise electricity, electrify everything' – then the far right serves as the most assertive defender of incumbent interests *at each step*. Since half of the world's oil is combusted in road transport, a hypothetical full and rapid electrification of automobility would be felt by the fossil fuel industry as an amputation of its lower body, with stranded assets reverberating through the elaborately interwoven financial architecture of New York, London, Toronto and Frankfurt. Mitigation in the transport sector requires reducing travel, shifting to active and public transit, and rapid electrification. What we find in this study of far-right movements and parties across North America and Europe is a near-universal obstructionism on each of these three fronts. The far right takes up a war of position at every turn on the road to decarbonisation.

Paths forward

In what comes next, we examine the rise of the far right and its battle against the green transition, with a focus on its politicisation of automobility and the internal combustion engine. Our study of far-right movements and parties centres on the European and North American context, where we home in on the most novel and escalatory elements of the conjuncture in these regions. In Chapter 1, we develop a theoretical account of the far right's 'inverted crisis' as a convergence of moral panics, thematise its relation to the 'organic crisis' as well as other background conditions, and consider its potential as a 'fascism-inducing crisis' moving forward. In Chapter 2, we track the political economy and ecology of the automobile, offering a materialist history of the combustion engine while exploring the moral economies on which the 'war on cars' plays out.

Following a roughly chronological order beginning in the pandemic, the remaining chapters trace some of the core actors and developments in the late far-right phase of denial. In Chapter 3, we focus on mounting opposition to traffic reduction schemes in the UK, which fused climate lockdown conspiracism with moral panics around climate protesters, traditional sex and gender roles, and migration. Throughout, we illustrate how opposition to 'green' urbanist schemes took a dialectic of radicalisation between street and electoral politics, running through resentful suburban geographies and culminating in the meteoric rise of Nigel Farage's Reform Party over the traditional Tory right and in massive riots and rallies like the Unite the Kingdom demonstration in London. Chapter 4 probes the use of the combustion engine as a protest tactic of the fossilised petty bourgeoisie, from Canadian 'freedom' convoys to European farmers' tractor protests between 2019 and 2024. Both protest cycles further entrenched carbon-intensive systems while augmenting a moral panic cycle centred on the food and wellness culture wars. At the extreme end, this moral panic cycle culminates in a micro-fascist body politics of carnivorous raw-food diets and extreme fitness regimens pitted against 'globalist-mandated' insect-based diets.

From there, Chapter 5 turns to the process of fossil fascisation in Trump's third presidential campaign, focusing on his apocalyptic rhetoric conjoining the EV transition to an immigrant 'bloodbath'. Following Trumpism back into the White House, we show how these crisis narratives were transformed from a rhetorical emergency to a state of emergency, with 'unitary' executive action dismantling climate regulations across the board. Finally, in Chapter 6, we examine the hardening of a mobility regime increasingly modelled on the concept of 'remigration', linking the expansion of militarisation, the crisis of the German auto industry, and the environmental racism of the far right. Radicalising inward and outward, ethnonationalist leaders use the inverted crisis to reinforce a 'lifeboat ethics' justifying their dual war on climate and migration.

As we progress through the chapters, we scale up in both time and space, from local protests against planning policy, to national and transnational mobilisations against environmental regulations, to the official politics of presidential campaigns and executive powers, to a global regime of mobility in which anti-migration and militarisation increasingly transform the world system in the image of fossil fascism. Taken as a whole, these multi-scalar cases illustrate the progression of numerous interacting cycles of radicalisation. Through affect, discourse and ideology, moral panics emerge and converge, attracting new layers of society into the far-right's conspiratorial orbit. Protests and convoys radicalise; anti-urban rallies transform into racist rioting; mainstream parties absorb extreme-fringe framings and yet are outflanked by a surging far right, which approaches high office through variants of Trump-ism and transforms global dynamics in the process. At each turn of this great driving right show, the far right is fuelled by crises of its own making.

1
The Inverted Crisis

In his essay 'The Great Moving Right Show', Stuart Hall showed how the crisis of Keynesian social democracy in Britain was exploited by Margaret Thatcher and her allies to build a new ideological consensus around neoliberalism. Thatcherism, he wrote nearly fifty years ago, sought to seize hegemony by proposing a radical break that 'takes the elements which are already constructed into place, dismantles them, reconstitutes them into a new logic, and articulates the space in a new way, polarising it to the right'.[1] Such ideological developments, Hall observed, may pre-exist or lag behind economic developments, or follow a different tempo altogether. In this case, Thatcher ideologically intervened by assailing the common sense of state welfarism and by building on a series of moral panics. Her 'respectable' right wing project absorbed the momentum of an extreme right embodied by Enoch Powell and the National Front. In its hegemonic struggle, Thatcherism could at once represent itself as 'moderate' vis-à-vis the 'fire-and-blood' racism of Powellism and nonetheless make effective use of its many moral panics – its warnings of 'social anarchy', attacks on racialised 'enemies of the nation' and calls for 'law and order'.

To diagnose the rise of the right, Hall followed Antonio Gramsci in distinguishing between *organic* and *conjunctural* historical movements. Whereas the concept of an 'organic crisis' refers to deep-seated structural contradictions of capitalism that are often decades in the making, the category of the 'conjunctural' indicates a more immediate terrain of struggle on which ruling forces seek to 'conserve and defend the existing structure' and to cure its contradictions 'within certain limits'.[2] The struggle between ruling and oppositional actors to resolve or overcome such a crisis is key to securing long-term hegemony.

Thatcher's project could thus be understood as a conjunctural response to an organic crisis. In the 1970s, the crisis stemmed from the contradictions of postwar social democracy, the failures of Keynesian economic management, overproduction, a falling rate of profit and the OPEC crisis. This constellation yielded a period of high unemployment, economic stagnation and rising inflation – or 'stagflation', a term coined by Conservative MP Iain Macleod in a speech attacking the Labour government. Although Labour had long represented the 'natural' party of crisis management, the Conservatives seized on the crisis of hegemony and fundamentally altered the balance of forces.

Today, we are again speaking the language of hegemonic crisis. In the long shadow of a global financial collapse, political earthquakes from Brexit to Bolsonaro, ongoing climate collapse, the coronavirus pandemic, Putin's invasion of Ukraine and Netanyahu's live-streamed genocide in Gaza, the ground has shifted beneath what recently counted as 'common sense'. Material symptoms abound, each a sign of deep structural malaise: stagnant wages, high energy costs, explosive housing prices, monumental wealth inequality, timid growth. At the beginning of his second term, Trump amplified the geopolitical turbulence with emergency orders, imperial posturing and sweeping tariffs. As establishment parties and coalitions continue to implode, the far right makes steady gains from one country to the next, turning fossil fascism into more than a hypothetical threat on the horizon.

The grand backdrop to these political-economic ruptures is planetary destruction and warming temperatures, which, without immediate and unprecedented collective action, will yield a world to make current calamities appear like calm days. While many now conceptualise the proliferation of interrelated crises in terms of 'polycrisis', we theorise these developments – and foreground the climate and biodiversity crises within them – through the Gramscian and Hallian notion of 'organic crisis'. The rise of the far right must be understood in relation to not only political-economic and social components, but also to the ecological dimensions of an accelerating organic crisis in capitalist (re)production.

Amid the ongoing hegemonic breakdown, the right's strategy of denial and delay cannot only take a defensive character. It must turn offensive. In this context, contesting common sense on climate change – and, crucially, the state spending and economic planning necessary to address it – is among the far right's central goals. It is to this end that its most advanced, reactionary and conspiratorial elements develop new strategies to advance an ever-expanding series of moral panics. As they multiply and converge, the most significant is what we call it.

The inverted crisis is the far right's conjunctural response to an emergent organic crisis, including and especially its relation to climate change. For the far right, the crisis is not climate change, but what nefarious elites and hostile 'others' will do in its name. Linking this 'threat' to an ever-growing number of moral panics, the far right deploys what we call strategies of inversion. These strategies, also explored in subsequent chapters, are structured around familiar and novel oppositions: realist versus alarmist, workaday motorist versus climate activist, rural versus urban, producer versus parasite. The far right relies on these strategies to generate an ever-growing sense of crisis. As a core logic in ongoing processes of radicalisation, the inverted crisis is a driving force in the rise of fossil fascism today.

The organic crisis

While the impacts of 'extreme weather events' increasingly dominate the front pages of tabloids and broadsheets alike, new scientific findings on the impacts of climate change are often buried under more 'relevant' headlines – if they appear in news coverage at all. Many such findings pushed to the margins of public attention address future economic impacts. If temperatures rise 2°C above pre-industrial levels – with some reports suggesting that this threshold will be crossed in just a few years' time – severe economic contraction is forecasted. A growing body of literature predicts trillions in adaptation costs and other climate impacts, declines in global average incomes, and drops in GDP of up to 25 per cent by mid-century and heading towards the next.[3] Overestimating the powers of estimation, however, remains a danger when it comes to capitalism's ability to find new sources of profit even amidst the wreckage.[4]

What often drops out of public debate, where debate exists at all, is that scholarship has also made crystal clear who the culprits are: the fossil fuel industry and the ultrawealthy. Over 75 per cent of greenhouse gas and nearly 90 per cent of carbon dioxide emissions stem from the fossil fuel industry, while two-thirds of global heating comes from the richest 10 per cent.[5] The roots of the organic crisis have a clear class character.

At the end of the 1970s, data revealing the present and future trajectory of global heating was typically more available to oil industry executives and high-ranking governmental officials, less so to scholars of cultural studies or political economy. Unsurprisingly, then, the conceptualisation of an organic crisis by Hall, and by Gramsci before him, left out an ecological dimension that now seems a natural part of the analytic.[6] It is not simply that growth steadily raises global temperatures, and that the impacts of warming temperatures in turn affect the prospects for growth. It is that the intertwined political-economic, sociocultural and ecological dimensions of contemporary crises are structured by capitalism's endless drive towards profit and accumulation. The concept of organic crisis needs to be reworked to incorporate these dimensions, while also foregrounding the ecological component within socio-economic reproduction. Together with theorists like Ståle Holgersen, we believe it crucial to redress this conceptual blind spot.[7]

As opposed to the concept of polycrisis – which correctly but vaguely posits a contemporary and continuous interconnection between various crises – an alternative formulation of 'organic crisis' gives greater primacy to political ecology, a firmer material basis in political economy, and a sharper sense of temporality. The budding organic crisis is, after all, a '*longue durée* crisis'. As eco-Marxist scholarship has long emphasised, capitalism's drive for accumulation is destructive of natural systems. Though centuries in the making, the effects of climate change now appear not just in continuous natural disasters, but in economic, political and cultural developments. Much more than just an arbitrary interconnection, these 'crises not only coincide but also feed into each other, shaping and possibly reinforcing each other', as Holgersen observes.[8]

To conceptualise the interlocking nature of the economic and ecological in a more concrete way, consider insurance. As traditional insurance companies increasingly leave high-risk markets to avoid major losses from climate disasters, new systemic vulnerabilities and opportunities come into view. After a natural disaster, such as the Los Angeles wildfires in 2024, insurers sometimes claim that a third party – say, a utility company – was responsible for some losses. But since pursuing these claims is costly and uncertain, insurers increasingly sell the rights to the claim to hedge funds, which specialise in risky investments. Hedge funds might lose money, or they might reap massive returns if the claim pays off. The worsening climatic conditions thus shift risk from the heavily regulated insurance sector to the less regulated world of hedge funds and speculative finance, and possibly to other pillars of the economic system like reinsurance companies and the state. In an increasingly uninsurable world, disaster capitalists will still roam free and plunder, to our collective peril.[9]

What such feedback loops reveal is the growing danger of cascading crises of systemic proportions. In a 2015 speech on the 'Tragedy of the Horizon', Mark Carney, as governor of the Bank of England, warned of the need to prevent a 'climate Minsky moment'. The idea references the work of American heterodox economist Hyman Minsky, who stressed the inherent instability of financial markets, whereby fragility begets fragility, eventually leading to a tipping point or 'moment' of collapse. As the 1970s confirmed, financial systems can look stable right until they are overstretched and then break down. Scholars have subsequently developed the concept of a 'climate Minsky moment' to indicate a time when the financial system is impacted by a natural disaster and the resulting losses could push the weakest players over the edge, resulting in a 2008-style crisis of a climatic variety. The ever-louder warnings issued by scholars and central bankers alike suggest that this is yet another *longue durée* tendency building within the organic crisis of our time.

Perhaps it is no coincidence – and no small irony – that both financial actors and central bankers invoke this concept when assessing the vulnerability of their portfolios and funds to climate change. In a parallel that mirrors the distinction between conjunctural and organic movements, these actors now frequently distinguish between 'transitional' and 'physical' risks. Transitional risk refers to the exposure of vested interests to a *political transition* away from fossil fuels: the liquidation of fossil fuel assets, phase-out of polluting technologies or a blanket ban on carbon commodities. Physical risks, by contrast, indicate the potential financial losses initiated by *climatic events* themselves: a drought impacting microchip production, thereby disrupting supply chains, or a hurricane halting offshore oil operations. Calculated from the perspective of asset owners, particularly fossil capital, the categories of 'transitional' and 'physical' point to different sets of events, temporalities and actors that exert influence on the dynamics and experience of risk.

Viewed through this schema, the concepts of the 'physical' and 'adaptation' correspond to the *organic*, while 'transitional' and 'mitigation' align with the *conjunctural*. Much like the organic crisis, the physical risks generated by global heating are incurred through processes internal to capitalist development itself. The logics of capital accumulation drive an increase in carbon dioxide and material throughput, resulting in rising temperatures, biophysical devastation and extreme weather events. The actualisation of transitional risks, on the other hand, is more a matter of politics: an outcome of the balance of forces – activist momentum that is translated into emissions cuts or corporate strategies to slow-walk or stall out. Here we are operating on the conjunctural terrain, always contested and often contradictory. While the organic dimensions of the climate crisis will only increase over time, it remains the case that, in the present, the transitional risks of the climate crisis vastly outweigh the physical in most places. For those with vested interests in fossil fuels, the conjunctural appears more urgent than the organic.

Time and again, states serve as crisis managers of last resort. In the near future they will be compelled to act 'in the last instance' to absorb the devastating shocks caused by their own failure, albeit years earlier, to discipline fossil capital and decarbonise economies.

From a long-term perspective, then, an organic crisis looms large for the capitalist state itself, which is faced with a number of increasingly impossible tasks. The broader structural dilemma is not new, but stems from capitalist state imperatives that came to a series of contradictions in the 1970s.[10] Demands on the fiscal state have only compounded in recent years, which at the same time has led many governments to tie their own hands on climate action by committing themselves to 'derisking' and guaranteeing profits for asset managers like BlackRock and Blackstone.[11] As states 'manage' structural crises instead of acting to prevent them, extreme weather events are likely to translate into climate-induced bailouts. If the question is 'to whose benefit', the answer and disastrous consequences are unfortunately predictable: governments defaulting on their debts and a sweeping push for austerity measures that socialise losses and privatise gains.

With the contradictions embedded in the imperative to decarbonise, the predicaments run deeper still. In attempting to force the shift away from fossil fuels – deploying bans, phase-outs and policies to accelerate the transition – the decarbonisation state runs up against the structural limits of its own form.[12] Fossil fuel sectors still sustain state revenues and employment, meaning that any rapid phase-out risks eroding the fiscal basis of state power, while legitimation in capitalist societies – historically tied to rising living standards and cheap energy – is threatened as the state intervenes more directly in the circuits of everyday life. Risking undermining the growth and legitimacy functions that sustain it, the state thus occupies an almost impossible position: obliged to decarbonise yet dependent on the very structures it must dismantle.

The question, in turn, is how to conceptualise these structural tensions *politically*. 'In capitalist societies', Holgersen notes, economic or ecological crises do not automatically translate into political crises because 'the state apparatus is much more resistant than what is often imagined.'[13] Gramsci made a similar point, noting how, at moments of crisis, hegemonic actors can successfully partake 'in organizing greater forces loyal to the regime than the depth of the crisis might lead one to suppose'.[14] Today an essential dynamic to this political organisation is a moral panic cycle that constructs ever-new crises for the far right to exploit.

Demands to decarbonise occur within constrained and stagnating economies, with nationalist and denialist political parties on the rise. The shifts in patterns of production for renewables or batteries further intensify geopolitical tensions and heightened militarism. The dissemination of climate denial and conspiracism occurs amidst political disaffection, lack of representation, bursting inequality and climatic breakdown. While the dramatic political action required to avert ecological collapse is forestalled, morbid symptoms will continue to appear in improbable places.

The moral panic cycle

The experience and implications of climate change have been inscribed into 'common sense', in the Gramscian sense of the term *senso comune*: a largely uncritical, partly unconscious way of perceiving the world that becomes 'common' or 'popular'.[15] This not only shows in public opinion polls, which, outside the United States, indicate that the crudest expressions of climate change denial have waned – thereby necessitating what we are calling a new phase of denialism. It is also reflected in the official register of governance and political economy: the prominence of the annual COP meetings and IPCC reports, the rise of insufficient ESG (environmental, social and governance) investing metrics, the practice of corporate greenwashing, the countless social struggles that pressure institutional actors to divest. Yet any such absorption into a 'common sense' is partial and incomplete. In most respects, the climate crisis has hardly been registered in our socio-economic practices and arrangements.

In a similar key to 'common sense', some scholars have referred to this deep social backdrop as a *moral economy*.[16] Drawing from this literature, we use the concept of moral economy to mean the legitimating norms and expectations that underpin particular economic practices and structure various social relationships, such as those between state and citizen, capital and labour, creditor and debtor, landlord and tenant, buyer and seller, or producer and user.

Automobility – and mobility writ large – can be understood in exactly these terms: a set of structuring practices with underlying moral codes. Made possible by the state, and often essential for meeting everyday social needs, the practice of driving has common sense expectations attached to it. From the 'Fordist' to the 'post-Fordist' era, auto dependency has remained a unique feature of the state–citizen relation. Car ownership has thus been linked to the real experience and imaginative expectations of rising material standards of living and norms of citizenship. At the same time, cars are embedded in a network of economic practices, integral to a broader 'system of provision'. Travel to work, school, the doctor, shops, gyms or family often belongs to the realm of necessity – what some describe as 'forced mobility', given the lack of alternatives – while other journeys are leisure pursuits, often viewed as a compensatory counterpart for time spent at work. Intertwined with the social structures of gender, race, class and age, automobility has also become a site of intense psychic identification.[17]

The concept of moral economy bears on three features of far-right political ecology and automobility: first, the classic populist antagonism between the immoral corrupt elite (the state) and the morally innocent majority (the drivers); second, the far right's ability to moralise the politics of climate – transforming debates over transition and sustainability into questions of hypocrisy and national virtues – that the left would otherwise like to put into largely economic or ecological terms; and third, the ways in which contestations of this moral economy between state and citizen (congestion pricing, low-emission zones, traffic reduction schemes and so on) provide legitimating grounds for the insurgent far right to challenge incumbent parties.

The budding organic crisis has the potential to disrupt existing moral economies of mobility, in turn ripening the soil for far-right moral panics. With housing, transport and other sectors in desperate condition following decades of neoliberal privatisation and under-investment, the state cannot guarantee the basic expectations of generations past. This has opened the space for right-wing framings of crisis – from the so-called 'war on cars', to the battles over bike lanes and street congestion, to the 'pothole politics' of decaying infrastructure. Just as Gramsci and Hall underscored that such crises call into question 'common sense', an organic crisis affects breakages in moral economies. Both 'organic' and 'conjunctural' events challenge prevailing norms and practices – or what counts as 'normal' – in an existing moral economy. Whether progressive or reactionary, conjunctural reframings of social conduct can – if successful – be scaled up into state conduct.

Breaches in a moral economy thus present an opportunity for the far right to generate social anxiety – to ensure that they are experienced by the public in the form of *moral panics*. The concept was first given its classic formulation in Stanley Cohen's 1972 book *Folk Devils and Moral Panics*, which examined the rebellion against the conservative nature of the postwar settlement in the conflict between the 'Mods and Rockers':

> Societies appear every now and then to be subject to periods of moral panic. A condition, an episode, person or group of persons emerges to

> become defined as a threat to societal values and interests; its nature is presented in a stylized and stereotypical fashion by the mass media; the moral barricades are manned by editors and bishops and politicians and other right-thinking people; socially accredited experts pronounce their diagnoses and solutions; ways of coping are evolved, or more often resorted to; the condition then disappears or submerges or deteriorates and becomes more visible. Sometimes the panic is passed over and forgotten, but at other times it has more serious and long-term repercussions and it might produce changes in legal and social policy or even in the way in which the societies conceive themselves.[18]

A few years later, Cohen's insight into the moralisation of disorder was expanded by Stuart Hall and his colleagues in *Policing the Crisis*. Shedding light on the construction of a moral panic around mugging, they demonstrated how such panics reflect deeper crises of authority, providing cultural material for an insurgent 'law-and-order' project.[19]

Drawing from these accounts, we could say that moral panics are a way of constructing a 'crisis', generating social anxiety around it, and producing scapegoats for it, sometimes with transformative consequences for society and state. The success of a moral panic often depends less on the plausibility of its claims than on the ability of those choreographing the alarm – whom we call *moral crusaders* – to gauge the pulse of society, displace attention from some deep-seated problems, and channel existing discontents into outrage focused on more favourable terrain.

As a vehicle for the far right to win legitimacy and secure hegemony, moral panics work through images, fantasies and anxieties. They both fuel and are fuelled by a sense that established ways of life – an existing moral economy – are being subverted by various 'internal enemies'. As Hall and his colleagues explained, 'It is as if each surge of social anxiety finds a temporary respite in the projection of fears onto and into certain compellingly anxiety-laden themes: in the discovery of demons, the identification of folk devils.'[20] Today, as we will see, these demons and folk devils span from 'woke' protesters to so-called 'globalists' and 'the deep state'.

Within or between particular conjunctures, moral panics may go through various phases, with one building upon another. In their early phase, Hall and his colleagues observed, moral panics aim to provoke official responses, inquiries, or statements from government authorities, which may or may not temporarily appease the moral crusaders that stoke them. Thereafter, in a middle period, they amplify and address themselves to a 'threat to society' imputed to various scapegoats and folk devils. Finally, in a later period, the moral panic expands to a 'general threat to law and order itself' – or, today, to invocations of 'invasion', 'insurrection' and related conspiracy theories about the 'Great Replacement' and the 'Great Reset'.[21]

Through processes of re-signification and amplification, moral panics thus change shape, re-channel social anxiety and take on new targets. As they spread and transform, each takes on its own tempo. Caught in the moral panic vortex, time itself seems to speed up. One panic may capture a particular news cycle, or a succession of them may recast entire periods of cultural discourse and social struggle. Weeks or months revolve around topics like 'cancel culture', 'freedom of speech', 'cultural Marxism', 'critical race theory', the 'climate cult', 'gender ideology', 'immigrant caravans', 'dog-eating refugees', climate protesters blocking the streets or the supposed threat posed by trans people. In discontinuous and non-linear fashion, some moral panics may have their own sub-cycles. The phantasmic attack on trans people, for example, has spanned panics over bathroom usage, gender-affirming care for children, drag queen library book readings and trans women in sports, among many other topics of projected anxiety.[22] The succession of moral panics also crosses borders, a key part of the far right's internationalisation. Consequentially, a moral panic over 'wokeness' has been exported and translated into new contexts, becoming a general or *global* conspiracy – 'wokeism', 'wokeismo', 'Wokeismus'.

Through online platforms, internationally minded moral crusaders thus push panics across languages and borders, which are used as interventions in new contexts and debates.[23] Moral panics are unevenly distributed not just between publics and contexts, but also across class – with the possibility of moving in multiple directions. The platform logic of Twitter/X, which contributed to the radicalisation of billionaires like Elon Musk and Bill Ackman, shows that these panics are not only a weapon of oligarchs, but also a potentially cross-class phenomenon of radicalisation. Even if Louis Althusser was wrong to claim that the ruling class *must* believe the content of its own ideology, this is clearly quite often the case.

The process through which a succession of moral panics emerges can be understood as a *moral panic cycle*. When one panic inevitably fails to 'resolve' underlying crises and loses its momentum, another often surges in its wake. From one to the next, moral panic waves do not emerge in a completely random pattern; rather, they are conditioned by underlying 'organic' developments as well as the various progressive initiatives to which they are reacting – on issues from climate justice and pro-Palestine campus activism to racist policing and transgender rights. In forging folk devils against progressive framings, there is often a collapse or merger between the far right's target: one panic is transposed onto different subjects, who are then made into the same or a similar kind of folk devil as the last. New chains of equivalence are formed, which together construct an increasingly generalised 'conspiracy'. This broadening and enjoining of different enemies – eventually accumulating into a *general conspiracy* – is the underlying logic driving the entire cycle.

This moral panic cycle – the movement through which individual panics accumulate into a 'general conspiracy' – is now a condition of the far right's ascension to power. Geared in an authoritarian direction, moral panics offer the far right an opportunity to intensify an already existing crisis of hegemony.[24] As the pendulum of hegemony swings between 'consent' and 'coercion', and consent is supposed to be its firmest basis, state violence and repression can still be justified as legitimate means of responding to 'crisis'. In Hall's time, a media-driven moral panic cycle helped to usher in Thatcher's restructuring of the state and to legitimise what he called 'authoritarian populism'. Today, the process not only involves the discursive, ideological and imaginative work of news media and right wing moral crusaders; it is also driven by the algorithmically mediated work of online platforms, the strategic work of interconnected right wing organisations and think tanks and the ever-deeper pockets of billionaire donors. At present, the moral panic cycle constitutes an internationalising form of far-right strategy and a driving force in a global process of fascisation.

In an era where far-right parties sit atop the polls – where fossil fascism is not a threat on the horizon but rising to power – moral panics become all the more central for denial, disavowal and delay. They are the far right's tool against the green transition and, more generally, against any and all of its perceived 'enemies'. In subsequent chapters we show how this authoritarian transformation takes place both 'from below' and 'from above', how it moves from increasingly fascistic campaign rhetoric into official state policy, and how it incorporates and mobilises seemingly disparate groups and demands. We will also explore how this process constitutes a defining feature of the far-right gambit for hegemony today. In a word, we will see how, both in and out of power, the far right aims to ensure that 'crisis' is the order of the day.

The inverted crisis

Amid an accelerating organic crisis – the interlocking of economic, political and climatic turbulence – the far right's strategy of denial, disavowal and delay turns from defensive to offensive. Through a succession of moral panics, this offensive and aggressive turn characterises not simply the extremist fringe but also far-right parties and governments. In the introduction, we named this phase of conspiratorial acceleration the fifth phase of denial. Building on the anti-ecological and anti-migrant elements of previous phases, fossil capital and the far right perceive oppositional threats as existential in nature: the 'life' of fossil fuels and Western civilisation is at stake. Executing an offensive counterstrategy not only means defending existing moral economies of mobility, then, but also contesting perceived progressive victories in the domain of cultural politics. It is from this hegemonic battle that a new conspiratorial architecture emerges: the inverted crisis.

The inverted crisis is the general articulation of the far right's moral panic cycle in this conjuncture. It refers to moments and movements through which specific moral panics converge into a 'general conspiracy' using strategies of inversion. Propelling the self-radicalising politics of the far right, it is used to channel social anxiety against the energy transition and against progressive momentum wherever it may appear. The inverted crisis is thus a core logic within ongoing processes of fascisation, and a driving force in the rise of fossil fascism today.

The concept of 'inverted crisis' can be used at different levels of abstraction, however, with both a narrower and a broader meaning. In the narrower sense, we use 'inverted crisis' to denote the far right's inversion of the climate crisis. In this framing, it is not climate change, but rather climate mitigation policies and their proponents that present an existential threat to society. In the broader usage of the term, the inverted crisis is the general form through which the moral panic cycle is articulated. It is the movement through which moral panics are merged into a *general conspiracy*. Climate protesters, pro-Palestine activists, drug cartels, terrorists, migrants, socialists and anti-fascists are all figures that can be linked or collapsed into seemingly equivalent elements of an overarching existential threat.

If the inverted crisis is a generalised form that far-right politics takes in this conjuncture, *inversion* is among its core strategies. As moral panics build into a cycle, inversion constitutes a technique that powers, structures and combines the individual waves. Inversion and moral panics are not only intimately related, but together are partially responsible for the shared sense that we are living in an 'upside down' world. And it is through individual moral panics that inversion scales up – from a discrete technique to a general strategy, from a discursive intervention to a structure of denial, from a refusal of one framing of political reality to a reverse construction of its own.

The strategy of inversion can be spotted just about everywhere. In Germany, for instance, Naomi Seibt, sponsored by the Alternative for Germany (AfD) and the US-based climate denialist Heartland Institute, styled herself the 'Anti-Greta', while anti-fascist and pro-refugee slogans were co-opted to defend the automobile: 'Kein SUV ist illegal' – 'No SUV is illegal' – read one AfD poster. Engaging in an exemplary mix of inversion and projection, tech billionaire Peter Thiel has called Greta the 'anti-Christ', which he defined as 'an evil king or tyrant or anti-messiah who appears in the end times'. In the UK, a coffin was carried at the front of a 2023 anti-ULEZ march on Downing Street, an appropriation of a symbol commonly used by XR to signify a funeral for the planet. Internationally, around the same time as some elements of the climate movement began to radicalise their tactics in the direction of sabotage – most evident with Les soulèvements de la terre in France – so too did the anti-climate vanguard turn to vandalism, targeting not fossil technologies, but the almost pathetically inadequate tools of the actually existing energy transition, like bollards and cameras. And as extreme weather events intensify, the far right puts an inverted twist on 'anthropogenic' climate change by claiming that various political enemies are the real cause – antifa and 'green' arsonists setting blazes, disaster relief funds misdirected towards migrants.

The spokesperson for the long-standing Canadian denialist organisation Friends of Science, Michelle Stirling, provides some insight into just how intentional many such uses of inversion are. When she was asked how to combat fifteen-minute cities and the climate 'zealots' behind them, she explained her response to these hysterical types:

> It becomes a real street theatre thing, and you have to be willing to confront that and copy what they do, even. I mean, I went down to City Hall in 2020 with a little sign in Swedish. And I did a video calling for 'Sundays for Climate Reason'. Just like Greta … So you know, you have to be willing to take their memes and flip them on them.

In the context of ecological crisis and an increasingly mainstream far right, the script is 'flipped', to use Stirling's words, and climate action comes to be seen as the source of creeping authoritarianism. Or as right-wing psychologist-turned-influencer Jordan Peterson put it in a Twitter poem,

> DIE you NetZero tyrants
> Leave us alone
> Leave our cars alone
> Leave our flights alone
> Leave our heating and fires alone
> …

> Stop forbidding the roads
> Back the hell off
> Or reap the whirlwind
> Go panic about the apocalypse by yourself
> In the dark.

Where climate change is often framed in apocalyptic terms, here the attempt to address it with net zero targets is represented as a more existential threat. At the 2025 UN General Assembly in New York, Trump offered a warning in the same register: 'Both the immigration and their suicidal energy ideas will be the death of Western Europe if something is not done immediately … This "climate change," it's the greatest con job ever perpetrated on the world, in my opinion … If you don't get away from this green scam, your country is going to fail.'

On the surface, all of this might appear as a form of trolling: hyperbolic pronouncements or acts of transgression that play well algorithmically, but have minimal effect in the 'real world'. Such inversions, however, are now at the heart of effective moral-panic-making. By existentialising the 'threat' of climate action and representing mitigation as tyrannical, the climate crisis is inverted into a crisis created by treasonous internal enemies to attack 'us all'. When generalised through the inverted crisis, the 'internal enemies' become ever more vague and yet interconnected: all can be declared illegal 'terrorist organisations' for the state to surveil, arrest and remove. The unsurprising trajectory of this process is radicalisation, both inward and outward: the abandonment of liberal democratic norms in pursuit of 'internal cleansing' and 'external expansion'.[25] This can be seen in Trump's calls to 'hang the Democrats', to extra-judicially bomb 'narco-terrorist' boats and kidnap heads of state, to seize new resources and territories, or to grant 'absolute immunity' to ICE officers who murder immigrants and protesters.

The strategy of inversion allows not only for ever-new chains of equivalence, but also for new connections between disparate topics. Pandemic lockdowns are said to usher in climate lockdowns; critical race theory is blamed for drag queens reading to children in libraries; a genocide occurring in Palestine is projected onto claims of 'anti-white racism' and 'white genocide' in South Africa. Even outside the so-called bubbles of social media, the world is said to be – and is often phenomenologically experienced as – upside down. Exceptional in this regard was the collective experience of the COVID-19 pandemic, the shock of which prepared the ground for some of the most twisted forms of constructing the 'irreality' of our times. In its wake, we are increasingly forced to navigate the distorted – or inverted – image of a world in crisis.

Among the many sources feeding the cycles of inverted moral-panic-making is nihilism. A deeply affective phenomenon characterised by meaninglessness, carelessness and futurelessness, nihilism is itself an effect of half a century of neoliberalism. If there is no such thing as society (or, we may add, ecological interconnectedness) – if life boils down to the incessant pursuit of human capital and shareholder value maximisation – then the erosion of shared meaning and experience is neither reducible to a so-called 'post-truth' moment inaugurated by Trump presidencies nor some disorienting turn in 'postmodern' philosophy before it.

Any account of this condition must also confront the torrents of disinformation and deceit long pumped through the pipelines of fossil capital as well as the resulting corrosion of trust in science over the same period of time. Neoliberalised nihilism and climate denialism are not only contemporaneous, but dual forms that feed into one another. In the new phase of denial, the far right's inversions are, to borrow from Naomi Klein, 'made thinkable by the baseline war on words and meaning in more liberal parts of our culture', beginning with an 'all-out war on meaning' that already crested in the previous stage of 'progressive-cloaked capitalism'.[26] When neoliberal greenwashing divorces the discourse of climate action from its substance, it establishes favourable epistemic terrain for the far right's denialist flight from meaning.

By now this has long been everyday 'reality': politics as branding; social media posts as social capital; language as cheap, disposable currency. In a world of constant lying and omnipresent conspiracism, which includes what Klein calls the 'Mirror World', our perception of events, others and ourselves is warped; it is shaped less by the fear than by the expectation that 'absolutely anything and anyone can be severed from their contexts and made to mean their precise opposite'.[27] We do not inhabit the same conditions as 1930s Europe. But if inversion is a defining feature of our time, and if anything can mean anything, we are again in a period in which 'everything is possible and nothing is true' – to paraphrase Hannah Arendt's account of the epistemic conditions fostered by mid-century fascism.[28]

In the current cultural landscape, nihilism isn't primarily about the loss of ultimate values. As Wendy Brown has emphasised, it is just as much about the reduction of politics to the pursuit of power for power's sake.[29] This becomes clear when far-right trolling and shitposting become a widespread social practice, when the manosphere conquers the podcast rankings and cartoonishly pathological forms of masculinity enter the core of youth culture. In this context, conspiracy thinking and inverted rhetoric resonates ever more widely and deeply. The COVID-19 pandemic mutated and accelerated these tendencies, blending together anti-establishment energies, QAnon communities, libertarian individualism, New Age wellness and anti-science ideologies. New forms of *diagonalism* disrupted traditional left–right distinctions through a conviction that all power is conspiracy. This ostensibly anti-corporate conviction was ready-made for grift-friendly movements like MAHA (Make America Healthy Again), which catapulted Robert F. Kennedy Jr into one of the most powerful public health positions in the world.[30]

In this post-pandemic phase of denial, technology rapidly blends the conspiracy theory smoothie. AI fills our feeds with slop and deepfakes to generate social anxiety and carbon emissions, with half a trillion in US federal spending running through AI data centres requiring a massive fossil-fuel-powered energy expansion. Increasingly, AI has become a signature form of far-right political aesthetics and surrealist propaganda production.[31] With at least half of society reportedly nervous about being replaced (in this case by robots, not immigrants), it is not for nothing – it is, in fact, eminently reasonable – that younger generations express despair and fatalism about the future.[32] While nihilism or fatalism may be prerequisites for a full embrace of AI political content, the technology itself tends to produce ever more nihilism and fatalism. All the while, the far right is caught in the gravitational pull of both techno-optimism and fascist apocalypticism.[33]

If 2015 was the year of birtherism, swiftly followed by the year of 'fake news' and 'alternative facts', we now inhabit the era of the Big Lie. The anti-democratic conspiracy theory, expressed as election denialism in its most extreme form, has become an increasingly international element of far-right platforms. The 6 January 2021 attack on the US Capitol building – to steal an election under the banner of 'Stop the steal' – stands out as a seminal case of inversion, as does the doppelganger movement backing Bolsonaro that descended on Brasilia a year later. Such grand-scale lies turned into far-right actions represent a peculiar form of untruth which, when backed by force, asserts itself as innocent and true and defines its opponent as dangerous and false. This aggressive form of denial aims to remake the world in its own image, materialising as potent firepower in the far right's project of ultra-national rebirth.

Fascism-inducing crises

There is, of course, a longer history of inversion in reactionary politics. As a counterrevolutionary tradition once tied to the *ancien regime*, conservatism itself has continuously adapted to different times and environments. In this process, scholars have observed, the right has developed 'substantive *antitheses* to progressive core concepts', such as reason and equality.[34] As the meaning of progressive ideas changes with historical context, so too do their conservative antitheses. 'The form conservatism takes', writes Edmund Neill, is 'parasitic on its progressive opponents'.[35] This parasitic quality, Michael Freeden similarly observes, generates a 'structural mirror-image reaction to whatever ideological configuration is perceived as the most immediate or menacing source of externally induced change'.[36] Emphasising how the right 'copies and learns' from its opponents, Corey Robin suggests that this has continuously led to an 'absorption of the ideas and tactics of the very revolution or reform it opposes'.[37] Right-wing reaction, we might say, tends to parasitically mimic the evolution of the host.

How does such mimicry manifest today? A central mechanism could be called 'inverted victimhood'. By parroting leftist critiques and struggles against social domination, right-wing movements have – sometimes consciously, sometimes unconsciously – reversed the roles of exploited and exploiter, oppressed and oppressor. In doing so, claims to inverted victimhood evolve vis-à-vis different progressive and liberatory struggles. 'In the Mirror World', Klein observes, 'there is a copycat story and answer for everything, often with very similar key words.'[38] In this new, post-pandemic phase of denial, such strategies of inversion have accelerated in speed and inflated in scope: the motorist is the victim of the climate activist; any solidarity with Palestine and all critiques of Israel are equated with anti-Semitism; and the label of the 'terrorist' is no longer primarily applied as an anti-Muslim slur, but capaciously used to justify the repression of anyone on 'the left'.

Historically, there are several famous cases of inversion being paired with semantic inflation: men are the victims of feminism, whites are the victims of racial-justice movements, bosses are the victims of labour struggle. At play today are the anti-ecological inversions deployed by the fossil fuel industry itself. A central instance can be found in the 'industrial apocalyptic' rhetoric used by the coal industry to co-opt powerful discursive frames developed by the environmental movement.[39] 'By repeatedly raising the spectre of job-killing regulations, energy industry annihilation, a backward slide into the "dark" ages of limited energy access, and widespread economic catastrophe', Jen Schneider and colleagues write, 'the coal industry and its surrogates make apocalyptic appeals a central part of their rhetorical strategy.'[40] In tying an impending 'apocalypse' to environmental regulation, pro-industry organisations arrived at a now-infamous slogan: the 'war on coal'. This apocalyptic rhetoric – which could be considered a forerunner to the inverted crisis – was a twist on the strategy of the Wise Use movement, a corporate-funded 'grassroots' campaign which sought to steal environmentalists' anti-establishment identity while forcing them to defend themselves against claims of being government insiders peddling special interests.[41] This instance of inversion has also been called *aggressive mimicry*, a term derived from biology for how predators disguise themselves, including how they disguise themselves *as prey*.[42] More colloquially, we might say that the 'war on coal' – like its cousin, the 'war on cars' – is a wolf in sheep's clothing.

Antonio Gramsci's account of fascism as a political strategy premised on 'aping' working-class struggles is instructive here. According to Gramsci, fascism was marked by a change in the forms of political expression of the petty bourgeoisie: a precarious middle stratum characterised by partial ownership of capital and a strong ideological alignment with self-reliance and autonomy. Having previously been parliamentary in character, with the rise of fascism this class 'changes the form of its working practice, becomes anti-parliamentary, and seeks to corrupt the streets'.[43] Fascism thus pivots away from parliamentarism on behalf of the petty bourgeoisie, who seek to imitate and 'corrupt' the form of political expression hitherto dominated by and associated with the revolutionary proletariat: mass mobilisation. Thus fascism could be said to operate through a mimetic mode of political expression, which 'apes the working class, and takes to the streets' – a political form at once imitative yet antithetical in substance.[44]

If inverted victimhood has a diverse and storied past, fascism elevates the sense of victimhood to crisis proportions. Robert O. Paxton's famously distilled definition of fascism begins by describing it as 'a form of political behavior marked by obsessive preoccupation with community decline, humiliation, or victimhood'.[45] In a largely congruent description of fascist ideology, Roger Griffin highlights its core myth: 'an organic "people" forming an "ultra-nation" is in crisis and needs to be saved from its present state of disintegration and decadence through the agency of a vanguard'.[46] Both accounts come down to the same fundamental myth: the nation is in crisis and it is up to a vanguard of patriots to restore national glory. Inverted victimhood works to construct such a crisis and to give a face to the enemy, the cause of national humiliation. Fascism is then presented as the vehicle of this existential struggle for redemption.

The tendency towards national crisis is, however, technically distinct from what the historian of fascism Geoff Eley calls 'fascist-producing crises', which denote such conditions ripe for fascist politics.[47] While often uncoupled from empirical reality, crisis narratives may gain or lose traction depending on the sociocultural, institutional or economic character of particular historical conjunctures. The experience of war and recession, for instance, offer particularly fertile soil in which crisis narratives can take root. The 'profound structural crisis' of interwar liberal democracy, too, was an important background condition for what was said to be the crisis facing the ultra-nation.[48] Griffin thus described the crisis of Weimar German democracy as generating a 'powerful palingenetic climate': a period in which the far right's core myth of national crisis and the necessity of rebirth (palingenesis) takes on a particularly powerful resonance.[49]

Today, in what we have called an accelerating organic crisis, a similarly palingenetic climate is taking hold. The pandemic, inflation, the cost-of-living crisis, increasingly intense and frequent extreme weather events, ongoing and new wars leading to death and displacement – all swirl together to form precisely such an environment. What historians of fascism have conceptualised as a fascist-producing crisis thus resonates with 'organic' conditions conducive to the rise of fossil fascism. Responding to these conditions, the inverted crisis is a conjunctural intervention through which the far right generates social anxiety and builds momentum for its disruptive form of hegemonic struggle.

The inverted crisis can thus be seen as today's climatic variation on fascism's inherent need for a sense of overwhelming national crisis beyond all traditional solutions: fascism's 'obsessive preoccupation with community decline, humiliation, or victimhood'.[50] In response to the relentless crisis narratives of their own making, fascists propose a palingenetic revolution: a rebirth of the racialised, masculinist nation. By actively cultivating crises, real and perceived, the far right thus rises through and not despite its contradictions. It both relies on constant crises and claims that it possesses the only solutions to transcend them. While it holds onto fossil modernity until its last dying breath, it tells its supporters it represents the only exit ramp.

2
The Motor of History

If the wealth of capitalist societies presents itself as an 'immense accumulation of commodities', then our investigation of the political ecology of far-right nationalism centres on the immense accumulation of that signature commodity of the past century: the car.[1] A microcosm of macroeconomic transformation and ecological destruction, the private automobile underwrote a historically unprecedented period of growth in the capitalist core, which accelerated material standards of living and forms of popular representation. The car was the modality through which post-war prosperity was lived – a material practice shaping everyday expectations and aspirations, instilled with ideas of freedom and class mobility. The car became both a driver of capitalist accumulation through spatial expansion and a bedrock of 'common sense' shaping norms of national identity.

Throughout the twentieth century, this commodity became charged with nationalist energies. The rise of consumerist and car-based national economies provided the foundation for a symbolic contract between the state and its citizens. The production of a 'people's car' became a staple of legitimacy for liberal and fascist states alike. As roadway infrastructure was built and alternative modes of transport were displaced, automobility was increasingly locked in as an entire way of life, part of what we have called a 'moral economy' undergirding the habits and expectations of citizenship. This was the case in the Fordist moral economy of welfare statism, with upwardly mobile segments aspiring to own homes and cars. Even with neoliberal state and industrial restructuring, and diminishing provisions of state welfare, the centrality of private transport persisted: citizens remained relatively insulated from fuel price increases and, even with constrained state budgets, the costs of auto dependency, both economic and ecological, stayed in the public sphere. The fortunes of domestic auto industries also served as a structural and subjective gauge for national economic vitality or decline. In this way, cars could be conceived as the sensuous appearance of crisis.

Today, cars are embroiled in the climate crisis and its associated geo-economic rivalries. Indeed, the resource- and energy-intensive patterns of automobility have been primary forces behind the climate crisis. Yet as states haltingly pursue the phase-out of internal combustion engines and Chinese electric vehicle manufacturers rise to dominance, cars sit at the centre of distributional conflicts over mobility and energy consumption. Since many regard states as the inevitable agents of decarbonisation, the already strained legitimacy supporting state–citizen relations is further tested. A truly effective programme of decarbonisation would fundamentally disrupt and change long-standing norms of mobility and the material expectations tied to them. If class compromise and state legitimacy were purchased through a moral economy predicated on fossil-based consumerism, it is no surprise that states cannot easily give up the latter without sacrificing the former. An attack on this particular commodity appears to many as an attack on the wealth of nations, not only violating paths of accumulation but striking at the contract of belonging between peoples and states.

Decarbonisation and the electrification of cars signify a challenge to an entire order of common sense. Such disruptions are mediated, and often distorted, through media-driven moral panics, producing fertile ground for far-right mobilisation. Auto nationalism, in this way, provides substance to contemporary nationalist framings of 'crisis', or what we have called the inverted crisis of the far right.

Manufacturing morals

In 1934, writing from his prison cell, Antonio Gramsci reflected on the sweeping transformations reshaping industrial production in the United States. The rise of the automobile and the wider Fordist factory, he argued, signalled a new economic and moral order, one that imposed discipline upon the worker, imbued the wage with ethical significance and blurred the line between production and private life. Of the industrial innovations underpinning this order, the assembly line and vertical integration were most significant. Ford had introduced the moving assembly line in 1913, subjecting workers to repeating the same easily learned task again and again – boosting productivity and disciplining labourers. Simultaneously, the direct management of inputs and distribution – from rubber plantations to railway tracks – permitted lower production costs, translating not only to lower prices but also to higher-than-average wages. After the invention of the internal combustion engine in 1885, the auto industry had comprised skilled labourers producing a few bespoke vehicles – affordable only to an upper stratum of society. Fordist industrial methods, however, made cars increasingly available and began to organise the 'whole life of the nation' around production.[2]

If 'the history of industrialism has always been a continuing struggle against the element of "animality" in man', as Gramsci posed, it is telling that Ford found his inspiration for worker discipline in industrial abattoirs.[3] Visiting slaughterhouses in Chicago and Cincinnati, he saw animal carcasses moving between butchers on overhead conveyor belts. Translating this logic to cars, Ford devised an assembly line where parts moved between stationary labourers, 'taking the work to the men rather than the men to the work'.[4] This echoed Frederick Winslow Taylor's idea of scientific management, which sought efficiency through control and standardisation of the labour process. Also intent on taming animal spirits, Taylor aimed to eliminate the 'human content' from work, leaving in its place what he infamously described as a 'trained gorilla'. With labour rendered interchangeable, it was not uncommon for fifty different languages to be spoken on Detroit's shop floors.[5] Migrants to the auto plants came mostly from the American south and European countries, with many also coming from the Middle East to Ford's Dearborn factories.

Fordism was never confined to the factory. The vertically integrated, rationalised production process was complemented by, in Gramsci's words, 'extremely subtle ideological and political propaganda' in which workers were not only compensated for their labour, but also their conduct: their temperance, thrift and domesticity. Ford and his inspectors in the company's Sociological Department monitored his employees' private lives – what they ate and drank and how they lived. Though eventually abandoned due to cost, this system of surveillance endeavoured to equate producers with virtue. More than a disciplined labourer, Ford desired a moral industrial citizen – a 'master in industrial method' who would volunteer 'for the mines and the railroads' should 'the fires of a hundred industries threaten to go out for lack of coal'. His ambition, then, was to 'mould the political, social, industrial and moral mass into a sound and shapely whole'.[6]

Such expectations were sustained by relatively high wages. Before the five-dollar day in 1914, the annual turnover of workers at Ford was so high the company had to recruit nearly a thousand employees to retain just a hundred. This new pay packet, judged a 'family wage', was the basis for the male-breadwinner model, or the 'Fordist family', accentuating the wage's 'moral dimension' while granting legitimacy and a certain degree of bourgeois respectability to sections of the working class.[7] It also supported emergent norms of working-class consumption.[8] 'I'm not saying our workers will sing Caruso or govern the state. No, we can leave such ravings to the European socialists. But the workers *will* buy automobiles,' Ford announced.[9] Separation from the means of production was redeemed through consumption; investing the working class more deeply in the relations of production.[10]

While ensuring control of inputs and distribution served to lower production costs, there was also an ideological motive behind Ford's pursuit of vertical integration: to limit his dependence on others.[11] Ford owned and financed his projects internally, as he 'loathed banks and outside investors and was determined to maintain total control of his company'.[12] His disdain for finance capital was inseparable from his full-throated anti-Semitism. Openly espousing Jewish hatred, Ford was 'a major organ of anti-Semitic conspiracy theories', distributing *The Protocols of the Elders of Zion* and *The International Jew*, a serialised diatribe of conspiracies, through his newspaper, the *Dearborn Independent*.[13] Throughout the 1920s and 1930s, the paper was circulated across the company's dealership network, with many dealers putting a copy of the *Independent* in the glove box of every new car.[14]

Ideologically, these practices and beliefs coalesced into what Stefan Link has called an 'industrial moral economy', wherein a community of producers struggled against financiers and idle profiteers. Ford praised the man who 'earned his bread' and rebuked those 'who gamble with the fruits of other men's labours'. His rhetoric echoed late nineteenth-century Midwestern populism, wherein the producer contributes to the nation's wealth while the rentier and financier siphon from it.[15] 'The negation of the industrial idea', wrote Ford, 'is the effort to make a profit out of speculation instead of out of work.'[16] The logic underpinning this moral economy, then, aligns with the core tenets of 'producerism': an ideology distinguishing those who contribute to the prosperity of their community and those who live off the resources that they did not produce – the virtuous maker versus the parasitic taker.[17]

The industrial moral economy of Fordism became especially attractive in Italy, Germany, Russia and Japan – all countries wishing to escape the grip of British and American financiers.[18] Throughout the 1930s, specialists from this 'transnational illiberal modernist network' visited Detroit, spending 'weeks, months, even years at River Rouge to learn the American secret of mass production'.[19] 'Only a single great man', to the chagrin of the Jews, 'still maintains full independence', Adolf Hitler wrote admiringly of Ford in *Mein Kampf*. He hung Ford's portrait in his office, and in 1938 the Nazi regime awarded him an official honorific: the Grand Cross of the German Eagle. Another senior Nazi on trial in Nuremberg – Baldur von Schirach – credited a collected volume, *The International Jew*, as the source of his bigotry.[20] The flowering of nationalist imaginations, it seems, often has cosmopolitan roots.

Both Hitler and Mussolini conceived grand projects for mass auto production with VW at Wolfsburg and Fiat at Mirafiori. For Hitler, cars were 'mankind's most marvellous means of transport'; his ambition was to build 'a people's car' that sold for under a thousand reichsmarks.[21] Originally it was called the 'Kraft durch Freude Wagen' ('strength-through-joy car'), later changed simply to 'Volkswagen'. For Minister of Propaganda Joseph Goebbels, Germans would be 'a happy people in a country full of blossoming beauty, traversed by the silver ribbons of wide roads, which are open to the modest car for the small man'.[22] This vision of modernity, oriented towards the future, never materialised in the way its architects foresaw. Limited by fuel and rubber shortages, and reliant on forced labour, not a single finished car reached a civilian as war consumed Europe and millions lost their savings in deposit schemes: the people's car was a 'disastrous flop'.[23] Yet some of the silver ribbons were built. Germany and Italy expanded motorway networks – the *Autobahn* and the *autostrada* – laying the foundations for the automobile's privileged position in this modernity.[24] When a delegation of British MPs inspected the newly minted *Autobahnen* before the war, the team were dazzled by the 'sheer clean beauty' and 'vigorous sweeping curves' of the road, which seemed to conform to the natural landscapes: a blueprint to follow.[25]

On the eve of the Second World War, a transnational template for modernity had been set. According to Link, exchanges throughout the 1930s 'laid the groundwork for the infrastructure of global Fordism'.[26] Yet it was the shift to a war footing that accelerated its spread: the principles of mass production were embraced in the armaments factories of Nazi Germany and the Soviet Union alike, even though the latter never developed a substantial automobile industry.[27] Volkswagen's factory in Wolfsburg, built partially with confiscated trade union funds, shifted to military production, and was preserved by the Allies seeking to rebuild West Germany in the war's aftermath. Meanwhile, in the United States, the coming of war and the promise of lucrative federal contracts dramatically reoriented the relationship between industrial capital and the New Deal state. Detroit, temporarily suspending its identity as Motor City, was rechristened the 'Arsenal of Democracy' by Roosevelt. The carmakers now 'built tanks, airplanes, radar units, field kitchens, amphibious vehicles, jeeps, bombsights, and bullets – billions and billions of bullets'.[28]

The industry of industries

In 1946, business analyst Peter Drucker identified auto manufacturing as the pinnacle of 'modern industry'. It is 'to the twentieth century what the Lancashire cotton mills were to the early nineteenth century: the industry of industries'.[29] We can take from this a double meaning: the car industry was not only a prototype for the wider industrial economy, given that the production of this one commodity was emblematic of wider shifts in economic organisation. It was also its pacesetter, with the car dictating the tempo for associated industries. The formulation 'industry of industries' thus draws attention to the raw materials the car is made from and the myriad economic practices it enables, situating the car as the coordinating node for expanding quantities of matter and energy, a lynchpin in wider productive interrelations.

From this perspective, automobility can be seen as a central driver widening the 'metabolic rift' a term used describe the rupture between humanity and its natural conditions.[30] With humanity's ecological interchange with nature subsumed and directed by capital's endless circuits of accumulation, the exploitation of natural resources and the aggregation of waste ensues. Upstream, the automotive industry exists in productive symbiosis with steel, oil, glass, rubber, aggregates, concrete, electronics, engineering and construction, each transforming flows of matter first to value and then to waste. Downstream, a radiating network of operations is set in motion after the car rolls off the assembly line – credit and insurance providers, marketing and sales showrooms, maintenance and repair workshops, petrol stations and car washes, distribution and logistics systems, the arterial networks of highways and roads, car-dependent retail parks and roadside restaurants – each with a distinct social and ecological metabolism. Capital accumulates through a spiralling network orbiting the car: activating numerous circuits, the system of automobility is an expression of self-expanding value in motion. Guided by the imperatives of capital accumulation, automobility deepens the metabolic rift on multiple fronts and prepares the ground for the 'Great Acceleration' of ecological destruction.[31] The car might well deserve the epithet of 'the Anthropocene's battering ram'.[32]

No industry's fortunes were more intimately tied to the car than oil. At the beginning of the twentieth century, oil markets were threatened by the transition away from kerosene and gas lighting towards electric light. But primitive fossil capital was rescued by the automobile.[33] Since their initial coming together, the oil and automotive industries have remained interlocked, the availability of oil influencing the character and direction of the car industry. When the east Texas oilfield was discovered in the interwar years, gasoline prices fell sharply, incentivising bigger engines and vehicles, while encouraging longer journeys. With fewer domestic discoveries, high import tariffs and fragmented markets that curtailed economies of scale in production, such extravagances were unavailable to Europeans, who became accustomed to smaller, more efficient cars. Indeed, the 'automobile revolution' was most advanced in the United States, which not only consumed more oil per vehicle, but also was home to the most vehicles in absolute terms, boasting four-fifths of the world's car fleet on the brink of the Second World War.

At the same time, the booming car industry shaped the oil industry. Seemingly, the only thing that could slow down the automobile takeover was diminishing fuel supplies, fear of which had become a fixation of American oil companies and government officials after the Armistice. The First World War had been decisive: oil and autos became central commodities tied to national security and imperial ambitions. Inaugurating a new era of industrialised warfare, navies replaced coal with oil, cavalry was supplemented with tanks, and horse-drawn transport gave way to motorised vehicles.[34] Not wishing to return to the wartime 'gasolineless Sundays', and with the US Geological Survey warning of an impending 'gasoline famine' stemming from the automobile's voracious demand, the head of that body, George Otis Smith, urged the US government to support domestic oil companies seeking overseas exploration. This was the dynamic that produced American oil's pivot to the Middle East.[35] Fittingly, a local agent of the Ford Motor Company intermediated the first negotiations in the early 1930s for Standard Oil of California's rights to Saudi oil.[36]

The forces of destruction accumulated through the Second World War. Soldiers were increasingly displaced by 'extraordinarily powerful machines' requiring 'growing quantities of raw materials and energy'.[37] The United States, Drucker wrote, indulged in a 'veritable orgy of natural-capital consumption', leading to soil depletion and overconsumption of 'irreplaceable fuels and ores'.[38] As wars became more deeply integrated into the industrial system, the relationship between society and nature was further brutalised. Tanks, manufactured in Allied and non-Allied auto factories alike, were a 'developmental model' for clear-cutters, harvesters, bulldozers and other 'brute force technologies' later used to level forests, buildings and land.[39] The demands of war vastly expanded productive infrastructure. The network of pipelines and refineries necessary for military airfields, for example, was foundational for the petrolisation of societies.[40] In short, military mobilisation was industrial mobilisation.

As automobility spread across society, writes Adam Hanieh, 'it was no longer just industry and the military that demanded ever-increasing quantities of petroleum: oil consumption was now propelled forward by the daily needs of the individual and the household'.[41] Beyond powering vehicles, the wider machinic complex also depended on fossil fuels. The infrastructure required to get oil from basin to fuel tank involves refineries, pipelines, tankers, storage facilities and service stations – demanding tonnes of coke-produced steel. If coal and oil infuse the finished product, they also bind together the hardware they run on: asphalt and cement, the core components of roads, are generally both dependent on coal. This union of fossil energy and physical resources is consummated in those regions building infrastructure for private mobility, which rapidly escalated after the Second World War.

The infrastructural state

The decades following the end of the war witnessed an unprecedented period of economic expansion, at the centre of which stood the car and its wider infrastructural complex. Under the Bretton Woods framework, states sought to lift limitations on the mobility of capital, commodities and labour – an ambition for which automobility was both a means and an end. One way to absorb surplus capital and excess military capacity, cars and their infrastructural hardware facilitated the movement of goods and services, the flow of imports and exports, the spatial reorganisation and social reproduction of labour. The auto-industrial complex became a privileged target of state planning and investment: the construction and continual expansion of road networks connecting newly built suburban housing; the deliberate neglect and widespread decommissioning of alternative modes such as trams and trains; the creation of favourable fiscal regimes, including generous tax breaks for company cars; subsidies for fuel; and the direct provision of grants and financial support to manufacturers themselves.[42]

This state-directed modernisation and territorial project was also fuelled through external funding: the Marshall Plan's European Recovery Fund allocated more resources to subsidising American oil imports than any other commodity, explicitly funding oil refineries and transport infrastructure. Through the 1950s and 1960s

Ford factories in Dagenham and Cologne, Volkswagen's Wolfsburg plant, Renault's Flins-sur-Seine facility and Fiat's Mirafiori complex rivalled Detroit in volume. Mass production was in its heyday: cars, homes, appliances and holidays were all subject to Fordist principles.[43] These decades cemented the automobile as the commodity par excellence, the pre-eminent symbol of modernity and individual freedom, a freedom which depended on vast investments and coordination by the state.

The sprawl of car-dependent suburbs, woven into the expanding infrastructural complexes, gained millions upon millions of new inhabitants in the postwar period. By 1960, half the US labour force now lived in the suburbs, surpassing city dwellers and rural residents. The car, alongside the home, became the principal unit around which working-class life was increasingly governed – the complementarity between the two fostering a 'gigantic expansion of commodities'.[44] These new social landscapes were not politically neutral. Already by the 1930s it was acknowledged that 'debt-encumbered homeowners do not go on strike', and with the car representing the most costly commodity bar the home, workers were co-opted into conservative politics through indebtedness and increased investments in economic stability.[45] The government encouraged homeownership through loans – the state paving the way to the suburbs in more ways than one.

The radical reorganisation of space was also a radical organisation of social relations. Automobility emerged as a condition for social mobility, privileging car owners and marginalising those without vehicles, a process stratified along lines of race and gender. Homes were crammed with consumer durables, restructuring domestic and labour rhythms while extending commutes and entrenching gendered divisions of care and work. Dissolving racial solidarities within the working class, suburbia emerged as the destination for 'white flight' from increasingly multicultural urban cores, with 'secessionist automobility' as the mode of separation.[46] The kind of freedom on offer, however, was one constantly plagued by delays, a promise deferred.

Already in 1960, Stuart Hall had touched on these contradictions of car-based growth in Britain: 'the car industry in the Midlands is booming: but the motor firms can scarcely get their lorries to the ports, because the roads are crowded, and the harbours are full and unreconstructed'.[47] With industry hindered by infrastructural constraints, the prevailing logic would be for the state to restore flow – to build another lane, another interchange. Yet the phenomenon of 'induced demand' suggests that, with new road capacity, the same problem would reassert itself on an expanded scale: after infrastructure is added and the initial reduction in congestion lowers journey times, more drivers are lured onto the road by the savings in time, meaning that traffic levels rise and the gains are eliminated. Despite extensive infrastructural investments in bypasses, orbital motorways, flyovers, tunnels and a constant churn of roadworks, congestion remains a persistent feature of road-based transport systems. In London today, the average speed of car travel is approximately twelve miles per hour – almost identical to a century ago.[48] Notwithstanding short-term fixes or partial geographic exceptions, as a general logic, as André Gorz argued in 1973, 'the car wastes more time than it saves and creates more distance than it overcomes'.[49] This is Jevons's paradox in infrastructural form. Building more roads to cure congestion is like loosening your belt to cure obesity, transport researchers like to say.

While this cyclical pattern is well established, state planners and policymakers routinely ignore it. The common explanation for this pursuit of expanded irrationality is the capture of the state by economic interests rooted in cars, oil and roads. Ralph Nader referred to the US variant as the 'Road Gang', pointing to the systematic decommissioning of trams across cities, orchestrated by General

Motors, while others have considered the US road lobby to be as significant as both the oil and armaments lobbies.[50] In the UK, a similar coalition of vested interests is said to be 'the single most powerful lobby', partially blamed for the removal of nearly half the country's railways.[51] Across the industrialised world, the road lobby has been 'spectacularly successful' in setting the course for state-directed infrastructural modernisation.[52]

Yet state capture is not the only way to understand this process. While deals were struck and alliances brokered behind closed doors, the car's embeddedness in an industrial matrix – its circuits of accumulation across auto, energy, construction, technology, finance, sales and steel – also bound the state to this trajectory, generating a momentum that seemed to exceed deliberate control. After all, capitalist states, much like individual firms, find themselves constrained by a dominating logic of capitalist accumulation in a competitive world system: 'Once accumulation is engaged upon it is not a choice, rational or otherwise' – states are now 'subject to a compulsion, severe and inescapable'.[53] For states, revenues and employment tied to the machinic complex, from manufacturers to insurance, dealerships to drivers, are not incidental, but integral to the reproduction of a capitalist system from which they cannot easily extricate themselves. More than state capture, it might better be understood as 'state dependence', with capitalist imperatives informing the 'decisions of state actors in key policy areas *even in the absence of outright lobbying*'.[54] Under such conditions, neither ecologically nor socially minded outcomes should be expected.

The nation mobilised

In debates over the role of the state, the nation is often conspicuous by its absence. Yet, as Claudia von Braunmuhl noted, 'The bourgeois *nation*-state is both historically and conceptually part of the capitalist mode of production.'[55] Nationalism, with its myriad cultural, political, economic, subjective and affective inflections, serves as the bridge between the 'nation' and 'state'.[56] Rather than something static and unchanging, nationalism is best conceived as a process that develops alongside the productive relations of capitalism: what it means to be a national subject changes in relation to shifting social conditions. As the car motors through history, then, the nation – its expectations and anxieties, its sentiments and symbols, its infrastructures and identities – goes along for the ride.

It was in the heyday of nationally organised mass production that the car's 'industrial imbrication with the national' was most pronounced.[57] The economic boom of the post-war years was centred around so-called 'national champions': British Leyland in the UK, Fiat in Italy, Renault in France, Seat in Spain, Volkswagen in Germany and Volvo in Sweden. The prominence of auto manufacturing in the national economy served as a symbolic 'indicator of modernity' and a 'significant measure of national economic and industrial virility'.[58] This was a period of 'excessively nationalistic emphases about science and technology', with such 'techno-nationalism' manifesting in a veneration of citizens as innovators or the belief that national peoples are most culturally aligned with new technology.[59] National imaginaries were also invested in the machines themselves. Writing in 1960, one *Guardian* journalist detailed the 'fiercely nationalistic' character of cars, both aesthetic and mechanical: A 'Volvo is unmistakably Scandinavian, a Pegaso is as fierce as a Castilian bull, as exciting as a flamenco dancer'; a Citroën has the look of a French 'country bistro'.[60] A Rolls-Royce captured the British 'native tradition of engineering excellence, craftsmanship and attention to detail'.[61] Queen Elizabeth owned five, each hood adorned with a George and the Dragon ornament in solid silver. (In 2025, when negotiating tariffs, Trump insisted these cars 'could not be made anywhere else'.)

But cars, and the post-war nation-building programmes they served, were only bound to a territory in appearance. The US provided machinery for the factories, primed the pistons with finance and funnelled oil towards Europe. Colonial assets and postcolonial exporters ensured that rubber, metals and oil reached shop floors. The Bretton Woods arrangement demanded that manufactured vehicles be primarily destined for export to earn foreign exchange. Labour in car factories was also sourced internationally: in Germany, the *Gastarbeiter* came from Yugoslavia and Turkey; in France, Algerians and Moroccans worked the shop floors; in the UK, there were Caribbean and South Asian workers; in Italy, Sicilians and other southerners headed north.[62] Despite these transnational coordinates, the car has remained 'freighted with national significance'.[63]

The nation was also inscribed in the infrastructure that cars depended on. Roads and the car's built environment physically and symbolically knitted together the nation, providing an infrastructural basis for national imagination.[64] As cars advanced, mobility was disembedded from particular localities, bringing the imagined community of the nation into view: after acquiring his first car in the 1960s, one man from Cornwall realised that it was now 'possible to go to the Lake District and Scotland, it was like going to a foreign land'.[65] More concretely, motorways, bridges, tunnels, service stations and bypasses helped forge the national territory as the 'privileged scale for social relations, economic development, political governance and affective allegiance'.[66] The practices shaped by such 'territorially integrated infrastructural complexes' fostered a form of banal nationalism – not as patriotic outburst or jingoistic excess, but as durable reminders of collective allegiance, which 'expressed and congealed a specifically modernist and nationalist preoccupation with societal homogenisation'.[67]

If the physical mobility offered by the car expanded the horizons of the nation, the attendant social mobility reshaped national expectations. The convergence between rapid capital accumulation, rising living standards and institutionalised welfare reconfigured the terms of citizenship. Enabled by higher productivity and cheap energy, the so-called 'post-war settlement' transformed the relationship between state and citizen – guaranteeing employment, rising wages and a swathe of welfare benefits – which can also be understood as a state-centred moral economy. Governments presided over the 'institutionalised negotiations between capital and labour', formally or informally.[68] This pact supported a virtuous cycle between the lot of the working class and capital accumulation, organised at the scale of the nation state. Through industrial relations and the welfare system, the working class was integrated into the capitalist system 'as capital sought to develop the productive forces without limit'.[69] Theodor Adorno put it more sharply: 'The proletarians today genuinely have more to lose than their chains, namely their small car or motorcycle as well.'[70]

The car industry drove the industrial boom in production and consumption alike, boasting the highest profit rates across national economies, while supporting accumulation in adjacent industries.[71] In 1965, as the US produced a record 11.1 million cars, one job in six was tied to the automobile industry.[72] This helped forge a particular image of the productive worker: typically male, formally employed and territorially anchored in the industry of infrastructure.[73] As Kristin Ross notes of France, it was in the post-war period that Renault workers 'came to serve as the incarnation of the working class' more broadly, and what was true of France held across most advanced capitalist states.[74] The car symbolised consumerist affluence and upward mobility. In the UK, 'the working class, the vast majority of the population in both 1950 and 1975, went from austerity to relative plenty,' with more than half its households purchasing cars throughout the 1960s.[75] In Sweden, the nation's vision of 'the people's home' was unimaginable without a car outside, and in this context 'a car for each man (yes, *man*) meant social justice'.[76] The average citizen in industrialised countries was able 'to live as only the very wealthy had lived in their parent's day'.[77]

Crucially, the auto nationalism of the post-war period was constructed as a rejection of both the discredited ultra-nationalisms that had launched the war, and the communist alternative said to threaten a new one. With the state playing an outsized role in national economies, dominant forms of nationalism were cast as productive, industrious, democratic, civic, liberal and mobile – mobility serving as both symbol and substance of national belonging. Haunted by the spectre of the USSR, consumption acquired a moral and political dimension: a bulwark in defence of freedom and a contribution to 'national prosperity and civic duty'.[78]

The borders of the nation state also locked in global hierarchies within the working class. Workers in the industrial North tended to receive far greater compensation than those in the global South. The border fixed 'customary living standards' at the national scale.[79] As Marx wrote, the cost of labour contains a 'historic and moral element', not calculable through calories and shelter alone. It must also be understood qualitatively in terms of the 'level of civilization attained by a country' and the 'habits and expectations' with which workers are formed.[80] Even if, as Étienne Balibar reminds us, national welfare statism was geographically and historically limited, it nevertheless cannot be entirely undone; rather, it constitutes a 'historically irreversible fact' whose influence still echoes today in global inequalities, uneven ecological exchange and welfare chauvinism.[81]

The axis of crisis

A common periodisation of the twentieth century delineates its Fordist and post-Fordist stages, the transition beginning in the late 1960s, accelerating through the 1970s, then consolidating in the

1980s. This was a period of overlapping crises in the capitalist core, with rising inflation, stagnating economies, fiscal strains, energy shocks, combative labour movements and emergent social movements challenging established consensus. The auto industry was embroiled in this compound crisis, causing a squeeze on profits that left its established firms teetering on the brink of collapse, requiring state bailouts and nationalisations. Structural fault lines of overcompetition and overcapacity were exaggerated further following the so-called OPEC oil crisis of 1973, which saw oil prices quadruple and the reliable flows of cheap energy dry up. For auto makers, these energy price spikes constrained productive capacities, forcing investments in fuel efficiency, while also curtailing consumer demand. With full employment and rising wages harder to sustain, there was an uptick in labour unrest, often finding its most militant foothold in the auto industry. If cars were the axis of the organic crisis of this period, they were also part of its solution.

The causes of the profit squeeze in the industry were numerous, but chief among them was intensifying international competition. The unproductiveness of Euro-Atlantic car plants had been laid bare by Japanese car makers, with their cheaper and more efficient cars rapidly encroaching on markets. Firms like Toyota, Honda and Nissan, producing cars of 'unmatched price and quality' adapted to consumer preferences, were now ready to 'mount their export offensive'. Competitive pressures in the industry were further accelerated by problems of overcapacity. Many of the productivity gains associated with Fordist production methods had begun to fizzle out by the end of the 1960s. Yet with vast sums of capital sunk into machinery and factories, firms could not easily abandon existing stock and had to continue operating at near full capacity simply to break even. As a result, overproduction remained a chronic condition of the industry: manufacturers produced ever more vehicles just to stay afloat, far exceeding what established markets could absorb.

In an attempt to assuage profitability concerns, European and American firms undertook 'industrial restructuring', adopting a 'wave of product and process innovations'.[82] Stockpiled inventories were discarded, labour markets loosened, vertical operations disintegrated and standardised models abandoned. With a seemingly endless range of fabrics, lacquers, leathers, sunroofs, rims and radios, by the 1980s no two identical cars left the Volkswagen assembly line in Wolfsburg on any given day.[83] But beneath these superficialities, diversity was in fact diminishing. Manufacturers maximised economies of scale through modular vehicle platforms – five-seat vehicles, fuel range of 250 miles, maximum speed over 100 mph – capturing as many customers in a single design base. This tireless pursuit of exchange value produced goods whose use value was desperately inefficient in terms of energy, resources and space.

The 1970s oil price spikes triggered demands for energy conservation, efficiency and 'independence' – the latter fed by racialised narratives casting oil-producing Arab states as villains. One expression of this shift was the Energy Policy and Conservation Act, passed by the US Congress in 1975, imposing limits on energy expenditure across industry. But the bill contained carve-outs and loopholes favourable to manufacturers. Corporate Average Fuel Economy standards, for example, mandated that fuel efficiency compliance is measured not by the performance of a single vehicle but rather across the average of a fleet. So, instead of counting fuel consumption by the model (say, the monstrous Chevrolet Suburban), efficiency levels are averaged out across all vehicles produced by one firm. Lobbyists also secured different standards for cars and trucks, creating exceptions for 'light trucks' and 'utility vehicles'. This 'SUV loophole' essentially incentivised manufacturers to build heavier, taller and less efficient cars and pass them off as trucks. The bulky vehicles were aggressively advertised to ordinary consumers: images of machines scaling mountains sold to suburbanites for the school run. Beyond the loophole, the SUV's higher price tag promised vastly greater profits for manufacturers than regular passenger cars, partly explaining why what started as a US phenomenon grew into a global one.

Despite escalating crude prices, states across the rich world sought to insulate motorists, decoupling retail petrol prices through subsidies. While the price of crude rose 680 per cent from 1973 to 1979, tax cuts limited retail petrol price increases to 116 per cent in the USA, 138 per cent in France, 226 per cent in the UK and only 32 per cent in West Germany. Despite motorists – especially US motorists – being protected from market turbulence, the discourse of 'pain at the pump' revealed their political power once leveraged.[84] With queues at petrol stations, strikes and trucker blockades, presidents Nixon and Ford were willing but unable to lift fuel price controls.[85] The turmoil helped secure fuel tax rises as 'the third rail of American politics'.[86] So extensive was state coddling of the motorist that, according to the IEA, 'in the long term, the oil shocks hardly dented the *remorseless rise* of fuel consumption by cars'.[87] Bucking the trend across most sectors of the economy, drivers throughout advanced capitalist nations sustained steady growth in fuel consumption throughout the oil price spikes.

Also applying pressure on state and industry was the 'rising cultural authority of ecology' pushing for reforms to the political economic order.[88] Whereas earlier resistance focused on the car's threat to pedestrians, attention now turned to pollution, material throughput and the social costs of car-based society. In 1980s West Germany, *Waldsterben* – forest death – triggered an environmental mobilisation that challenged auto supremacy for the first time proper. Images of blighted landscapes and slogans like 'First the forest dies, then the human' placed the car under growing scrutiny as a vector of environmental harm.[89] The federal government attempted to curb emissions without disrupting industry, introducing catalytic converters and unleaded petrol. German car makers nonetheless warned that up to 30,000 jobs were at risk – a refrain echoed in later decades. Catalytic converters became mandatory for all new cars by 1989 and, in 1993, EU-wide regulations followed. What began as a crisis of national ecology ultimately reshaped the political economy of automobility, forcing environmental legitimacy into the core of industrial production.

Auto capital globalised at ever greater speeds as it approached the twenty-first century.[90] Car manufacturers increasingly turned to geographical expansion as a profitability fix: opening new markets while exploiting cheap labour and longer working hours in poorer parts of the world. Already by the late 1960s, Volkswagen had opened several factories abroad – in Ecuador, Egypt, Mexico, Nigeria, Argentina, Peru, South Africa, Yugoslavia and three in Brazil – signalling the onset of the new international division of labour.[91] This transnational integration of the industry was closely linked to its capital intensiveness, creating substantial barriers to entry and exit, and encouraging concentration and centralisation. This drove mergers, alliances and joint ventures, reshaping the industry into a handful of powerful multinationals: cities like Tokyo and Detroit became hubs of strategic and financial command, even as production networks dispersed globally.

Emerging economies, drawn by promises of multiplier effects and high labour productivity, sought to emulate industrial strategies and build a social order around the automobile.[92] But unlike in the global North, where urbanisation and industrialisation generally advanced together, the South often urbanised without industry, leaving populations in informal and precarious conditions. Peripheral countries were not only sources of labour and materials for the production of cars driven in the core, but also crucial recipients of the metal cages once they no longer met the regulatory standards or aesthetic tastes of the global North. While in the 1960s ownership levels barely registered, by the turn of the century the South boasted roughly three tenths of the global car fleet. Car infrastructure locked in dependence and division: as one UN report observed, 'Urban sprawl and motorisation come hand in hand with the expansion of slums and gated communities.'[93] This is a social and ecological catastrophe rooted in the contradictions of competitive accumulation.

Crises, as Ståle Holgersen reminds us, are mechanisms for the transformative reconsolidations of the capitalist social order.[94] These turbulent decades presented exit ramps, where profound changes to patterns of consumption in the capitalist core might have been pursued.[95] Instead, the various crises consolidated the system of automobility – through overproduction, vehicle bulking, lenient regulations and geographical expansion – widening, deepening and entrenching the metabolic rift. The global fleet grew from 50 million in 1950 to 400 million in 1980, then doubling to almost 800 million in 2000 – transforming landscapes and lives alike. In *All That Is Solid Melts into Air*, Marshall Berman describes the motorscapes of modernity tearing New York apart in the name of progress. We might adapt his insight to automobility more broadly: 'It was destroying our world yet doing so in the name of values many embraced.'[96]

Deindustrialisation and its discontents

As production left the bounds of the nation in the capitalist core, the fracturing of car industries stoked national anxieties. From being Europe's largest car producer and world's top exporter in the 1950s, the experience of decline in the UK was particularly acute. The car industry peaked in the 1970s, then fell into 'rapid absolute decline'.[97] When Rover, the last major British car maker, was taken over by BMW in the 2000s, it triggered 'moral panics' about the end for competitiveness in a nationally fetishised industry.[98] For the British authors of *Rolls-Royce: The Best Car in the World*, the disappointment was palpable when the company was bought out by BMW: 'For the world's most elegant car to be built by a manufacturer of German economy cars comes as a shock.'[99] As with shipbuilding, steel and coal, the decline of the car industry severed another link to the nation's industrial past. While some continued to invest national pride in less mobile technologies – as Margaret Thatcher did with the M25 motorway around London, urging all to 'beat the drum for Britain all over the world' upon its opening – the broader experience of deindustrialisation was one narrated and politicised as a familiar story of national decline.[100]

Shifting patterns of production gave rise to new state ideologies. As full employment and universal welfare could no longer be sustained, governments across the West embraced programmes of market autonomy and entrepreneurial individualism.[101] In Britain, Thatcherism became the paradigmatic expression of this shift. Hall identified how Thatcherism framed Britain's decline as a moralised dichotomy, where 'social market values' of meritocratic competition and personal responsibility were being eroded by 'welfare scavengers', dependent on handouts from the state.[102] This anti-collectivist discourse affirmed the worth of the self-reliant citizen while stigmatising the perceived moral failings of others.

The punitive dimension of this transformed moral economy – alongside its integrative function – can be understood through a producerist lens: distinguishing those who contribute to the nation's wealth from those deemed to be draining it. Though producerism had been embedded in the rise of Fordism, particularly in the world view of the man himself, the moral economy of post-war Keynesianism had a more welfarist and collectivist reach. But the postwar compact had always been structured by exclusions of racialised minorities, women's reproductive labour, service or casual workers, welfare recipients and the unemployed. As the limits of this structure came under strain, the crisis of social democracy provided an opportunity to rearticulate such subjects as parasitic: the cause of the crisis.

While the symbolism of the car was aligned with anti-collectivism and self-reliance, many of the habits and expectations underpinning state–citizen relations around automobility persisted. Residual infrastructures fixed social relations and served as tangible reminders of a symbolic social contract with the state. States continued to prioritise private transport and its infrastructure, socialising costs to make up for the shortfalls in fuel taxes. These were foundational for legitimacy, sustaining everyday expectations around mobility: affordable fuel foremost. This was made apparent at the turn of the millennium, when oil price spikes triggered yet another wave of trucker and farmer protests, this time across Europe. In the UK, truckers protested fuel taxes denounced as 'green' impositions, and were widely championed in the press as productive, entrepreneurial citizens. Once again, claims to the nation served as a way of challenging disruptions to state-centred, fossilised moral economies.

Underwriting political, economic and ideological shifts was the 'new centrality of debt', which sustained high consumer spending despite stagnant wages, rising precarity and declining tax revenues.[103] States, in tandem with global financial institutions, pioneered mechanisms for expanded consumer credit, the deregulation of personal lending leading to an explosion in lease contracts specifically for cars. These developments were vital for auto makers, who needed to sell record levels of vehicles even when demand was flatlining, thereby externalising their own risk through the debt of consumers.[104] The landscapes of Fordism were also saturated in debt, with financial institutions in the US predatorily pushing mortgages on precarious and often racialised households in the car-dependent suburbs.[105] This road led to the 2008 sub-prime mortgage crisis, sending shockwaves through global markets and marking another watershed moment for the car industry.

In the US, car sales fell precipitously, down 40 per cent from 2007 to 2008, with employment plunging further. GM was pushed to the brink. Deemed 'too big to fail', the US state took 60 per cent ownership through its recovery fund, while overseeing workforce cuts, healthcare eliminations and pension curtailment.[106] In the aftermath of the crash, states poured hundreds of millions into new car subsidies – the Konjukturpaket II, Cash for Clunkers, Retire Your Ride, Prime à la casse, the Scrappage Scheme – to stimulate auto sectors and encourage upgrades to newer 'environmentally friendly' models. The following years saw a 'qualitative leap' in car finance, with instruments like 'personal contract purchases' helping ensure that manufacturers maintained new car sales to break even. With debt contracts now making up nearly 90 per cent of new purchases, and payments rising faster than wages, some fear that auto debt may trigger the next financial meltdown.[107] In effect, the making, selling and using of more cars was cast as a route out of the financial crisis, compounding the organic crisis further.

Cashing in on 'accumulated resentments of the neoliberal period', the far right emerged from the fringes, edging closer to the mainstream of advanced capitalist states.[108] As state budgets tightened under post-crisis austerity, conflicts over scarce resources intensified.[109] The Tea Party made hay on accusations of profligate spending after the financial crash, anti-state anger that Donald Trump later absorbed and channelled into rust belt states. In Britain, austerity paved the way for the anti-migrant scapegoating of Brexit, when the 'deindustrialised periphery ranged itself against the London establishment'.[110] Founded on the back of the financial crisis, the AfD pivoted from being a single-issue anti-EU party into a full-throated anti-immigrant one, with a stronghold in the deindustrialised east. As ecological concerns increasingly animated mainstream politics, the far right added climate denial to its repertoire, casting themselves as defenders of fossil-fuelled jobs, lifestyles and technologies – cars not least. As states impose limitations and move towards phase-outs on the internal combustion engine, nationalist imaginaries, often structured by nostalgia, are mobilised in its defence.

Building your dreams

In 1999, an *Economist* cover showed the Great Wall of China with a highway crossing over it. For some, the front page recalled the prophetic lines from *The Communist Manifesto*, where Marx and Engels declared that 'the cheap price of its commodities are the heavy artillery with which it batters down all Chinese walls'.[111] China was presented as a case of capitalist aspiration, tentatively entering the global capitalist order. At that time, there were roughly 6 million privately owned cars in China amid a growing demand for imported cars.[112] Fast-forward a quarter of a century, and the number of cars has multiplied by fifty and domestic manufacturers dominate new markets around the world.[113] For the Big Three of Detroit as well as Volkswagen, joint ventures with Chinese upstarts in the early 2000s were astoundingly profitable – a ticket into a billion-plus market of increasingly affluent consumers. But by the early 2020s, Euro-American firms were rapidly losing their footholds – Ford's sales collapsed, GM and VW were outpaced by the homegrown firms they mentored, and Jeep retreated from the country entirely. Driving the Chinese automobile sector's rapid rise is the relentless price war waged by domestic operators – firms able to produce cars for a little over a quarter of the cost of their European rivals. The dynamic in the *Economist*'s cover has reversed. With cheap Chinese commodities poised to conquer, tariffs are the fortification with which the West – particularly the US – protects itself. How did this situation come to be?

The Chinese state laid the foundations for auto dependency, overseeing an urbanisation boom, dizzying in both scale and speed.

Beginning in earnest in the early 2000s, rapid rural-to-urban migration was coupled with mass expansion of transport infrastructure. In the aftermath of the 2008 financial crisis, the process accelerated, with local governments transforming urban land and the built environment, attracting investors and accelerating infrastructural development.[114] As markets in the West reeled, surplus capital looked to emerging economies for lucrative returns; the infrastructural complexes for automobility in China proved to be a popular and profitable prospect.[115] David Harvey compared the Chinese boom to the postwar project of suburbanisation in the US, noting that both played the key role 'in stimulating the revival of global economic growth for a wide range of consumer goods, especially automobiles, but also cement, steel, and coal'.[116] While there is a similarity in the form of this infrastructural Keynesianism, the gulf in scale is staggering: one viral statistic revealed how in the two years between 2011 and 2013, China used more concrete than the USA did in the entire twentieth century.[117]

The government's *Made in China 2025* plan, launched in 2015, sought to upgrade the economy into high-value-added sectors, electric vehicles among them. Manufacturing was proclaimed to be the 'core of the national economy, the root on which the country is established, the tool for national invigoration, and the foundation for a strong country'. EVs had been gaining momentum in China since the early 2000s. The five-year plan in 2001, for example, had named EV and battery technology a priority science area. Then, in 2007, former Audi engineer Wan Gang became minister of science and technology, and is now widely credited with convincing Beijing to invest heavily in EVs. Municipal governments supported producers within their jurisdiction with grants, low-interest loans and easy access to land, while also purchasing the goods from these 'local champions' to use for electric taxi and bus fleets. This fragmented landscape of production meant that only the most productive firms emerged from their regions to compete on the national scale – a state-sanctioned and supercharged survival of the fittest.

Today, BYD, short for 'Build Your Dreams', stands as the crown jewel of this multi-decade project, the leading firm in a national sector with quadruple the capacity of the US, capable of satisfying half of global demand for automobiles. From humble origins as a battery manufacturer in the 1990s producing components for electronics and mobile phones, the company now stands at the pinnacle of the global industry, with the CEO of Ford publicly framing BYD as 'an existential threat'. It is perhaps ironic, then, that the principles of Fordism have made a comeback. Just like its Western rival Tesla, BYD embraces vertical integration, producing nearly three quarters of its products internally – including batteries, microchips and plastic parts. More than in-house production, BYD also owns lithium mines and several massive vehicle transporter vessels, including the world's largest, capable of carrying over 9,000 cars.[118] Exporting to new markets, they are also expanding production operations in Uzbekistan, Brazil, India, Turkey and beyond. Another hallmark of the company is highly labour-intensive operations, organised along Taylorist lines, with a workforce now approaching a million people (and counting). Given his rural origins, it is little surprise that BYD's owner, Wang Chuanfu, is often cast as the new Henry Ford.

In short order, Chinese auto makers emerged from the pandemic as leading global players. In 2025, they exported a world-leading 8 million cars (both EVs and combustion engine vehicles), an eightfold rise from just five years prior. Chinese firms are also expanding production operations in South East Asia, Latin America and beyond, particularly in EVs, with auto makers now investing more outside China than domestically, partially as a way to circumvent tariffs. This is part of a second, 'green', Belt and Road Initiative, which has seen a surge of investment led by the private sector, concentrated in solar, batteries and EVs, and said to be on a financial scale equivalent to the US Marshall Plan.[119] At the same time, the US has only intensified protectionism of its internal combustion sector and now risks becoming 'more isolated and less competitive' as developing countries become increasingly integrated with Chinese value chains.[120]

If it is sometimes said that the history of twentieth-century capitalism can be told through the car, a quarter into the twenty-first, this would not seem entirely misplaced either. For the first time in the history of capitalism, as Paolo Gerbaudo argues, the West is a technological laggard, and not just in any industry, but in the industry of industries.[121] From autonomous driving software to infotainment systems and, most significantly, on price, Chinese firms are racing ahead on every metric. One industry analyst predicted that if European and American manufacturers do not launch an affordable electric vehicle within the next five years, they may well be extinct by the 2030s.[122] Geopolitics, it seems, has become inseparable from the dynamics of the auto industry.

The tectonic shifts in auto manufacturing are producing symbolic and economic shockwaves. German manufacturers, long entangled in a mutually beneficial relationship with China's industrial rise, now find themselves gripped by a state of 'profound angst'. As one report put it, the country has started 'to realise that China's new automotive industrial base directly competes with Germany's manufacturing foundation'.[123] In 2024 BYD sold more electric and hybrid vehicles worldwide than all German car manufacturers combined.[124] The dithering with which German auto makers approached electrification reveals the clutch of techno-nationalism. As Wolfgang Münchau observes, in the German industrial imaginary, 'a fuel-driven car is a mechanical-engineering product' whereas 'an electric vehicle is a digital device at heart' and is thus 'offensive' to its own 'ideas of what constitutes a car'. The heirs of Rudolf Diesel simply could not look an EV in the eye. The identity forged in the diesel engine sparked an 'excessive sense of the industry's own power', leaving them blind to the electric juggernaut approaching in the rear-view mirror.[125]

If today nativism remains the core of the far right's appeal, it is underpinned by Fordist nostalgia and a vision of fossil-based reindustrialisation.[126] This appeal draws on more than a century's worth of nationalist entanglements with automobility. If automobility forms the material foundation for the reproduction of capitalist social relations, its curtailment is also the way in which crisis events manifest themselves and are experienced: fuel shortages, petrol price rises, limitations on use (imagined or real). Exploiting the contradictions of state-led decarbonisation, the far right situates themselves as the foremost defenders of the car and the lifestyles it supports. At the same time, the rise of China as an advanced industrial powerhouse lends credence to one of the far right's meta-narratives, that of Western civilisational decline. These are just some of the conditions to help grasp contemporary moral panics around decarbonisation – from low-emissions zones to ICE phase-outs, from fuel taxes to curtailed mobility – as rearticulations and defences of nationally bound moral economies, undermined by ecological imperatives and the competitive dynamics of capital accumulation.

3
Lockdown City

On 18 February 2023, around 2,000 people gathered in Oxford to protest against a proposed traffic filtering system that aimed to reduce car use. A twelve-year-old girl was handed the microphone to address the crowd. Doing her best inverse Greta, she yelled, 'How dare you steal my childhood and my future, and the future of our children, by enslaving us in your crazy digital surveillance prison.' Leaflets explained that the scheme was a plot to trick people into living in open-air prisons. Barbed wire fences, cameras watching every movement, a diet of insects required for survival: this was the 'new normal' smuggled in under green virtue. More understated, one placard summed up the paranoid mood: fifteen-minute cities were 'the thin end of a global wedge that will curb our freedom of movement'. Such concerns echoed in the House of Commons, where then Tory MP, now of Reform UK, Lee Anderson described the plans as 'an international socialist conspiracy'. Nigel Farage similarly warned, 'climate lockdowns are coming'.

Sustainable-traffic schemes – car-free zones, bike lanes, congestion pricing, low-emission zones and restrictions on polluting vehicles – have become battlegrounds in the war on cars. The perceived elite imposition of such measures condenses a broader sense of grievance shaped by decades of neoliberal restructuring, stagnation and the dislocation of traditional hierarchies. As atomisation rises and trust and cohesion decline across society, the car comes to signify self-reliance and autonomy in a system experienced as broken. When sustainable-transport schemes threaten that autonomy, they appear not as criticisable yet pragmatic responses to ecological crises but as a perceived loss of status and control and a fresh assault by hostile elites. Cohering on streets and screens, the backlash can be read as symbolic of the conjunctural convergence between ecological crisis and nationalist resentment. The far right helps organise and politically channel these affective forces.

Some of the far right's success stems from the uneven geographies of capitalist development, with suburban and semi-rural peripheries left behind the soaring asset wealth of the city. The multicultural metropole emerges as the object of status-laden comparison and resentment. Restrictions on car traffic are thus experienced through spatialised envy, a simmering resentment at the 'dethroned entitlements' of automobility.[1]

The pandemic, in its swirl of anxiety and disinformation, of naked authority and extraordinary measures, added to the forms of articulation opposing such schemes. If anti-immigrant xenophobia and Islamophobia were the *sine qua non* of the pre-pandemic far right, mounting opposition to public health mandates supplied fresh material for mobilisation. The border remained a site of contention, but with immigration halted by pandemic restrictions, the focus shifted from restricting the movement of nefarious Others to liberating 'the people' from internal constraints. Such injunctions merged with pre-existing suspicions of climate governance to produce a discursive mutation: the climate lockdown conspiracy. In this articulation of the 'inverted crisis', environmental policy and public health control collapsed into one tyrannical project of surveillance and restriction of individual freedom. A politics of resentment spiralled into a politics of paranoia.

Such discursive mutations spread widely. Drawn into Telegram channels, anti-lockdown protests and alternative media, many encountered a parallel political universe – complete with seasoned actors and ideological frames more totalising in diagnosis, more radical in remedies. 'Freedom' groups proliferated in this suffocating environment. Entrepreneurs became political entrepreneurs. Restaurant owners, nutritionists, wellness practitioners, preachers, life coaches, the self-employed, small-business owners and managers made common cause. Some emerged as social media influencers, podcast hosts, peddlers of alternative health and 'natural' products, anti-lockdown protest speakers or leaders of new organisations. In some cases, this meant sidelining or supplementing their business for the 'business' of the movement, further blurring the line between politics and grift on the far right.

This chapter examines the far right's opposition to sustainable transport through resentment and paranoia, tracing how anti-car and anti-urban mobilisations pair with reactionary politics of sex, gender and race. Such mobilisations gathered beneath an anti-authoritarian cloak – reframing decarbonisation as state overreach or green tyranny – while drawing on a repurposed language of freedom and resistance. Yet this diagonalist strand of anti-authoritarianism and anti-statism often transformed into the far right's own authoritarian prescriptions: the imposition of law and order, as states increasingly *police* (rather than attempt to mitigate) the climate crisis by ramping up repression against climate activists; the control of gender-non-conforming bodies as part and parcel of an aggressive defence of the fossilised family; and, finally, the targeting and removal of racialised elements in society via mass deportation.

While highlighting the international dimensions of these processes, our focus in this chapter is on the UK, and particularly on the radicalisation of both street-level and electoral politics that led to Reform UK overtaking the Conservative Party. We show how a mediatised feedback loop, partially set in motion by mainstream parties, legitimised and encouraged the growth of far-right grassroots movements. But these movements exceeded the control of the traditional right, who ended up playing catch-up to the forces they helped unleash. This culminated in the subsumption of the traditional right, outflanked and outgunned by Nigel Farage and his Reform party. The sequence roughly corresponds to what Stuart Hall called the dialectic between the 'radical respectable' and the 'radical rough' forces of the right.[2] Advancing through this dialectic, pro-car demos gave way to pogrom-style riots, leading up to the massive Unite the Kingdom rally in London, all erupting from the seething geographies of downwardly mobile resentment. As Reform absorbed the energies of street movements comprising the 'rough' forces, they now masqueraded as the 'respectable' forces of the right. Small boats and potholes emerged as twin symbols of 'Broken Britain'.

Climate lockdowns

The concept of the 'fifteen-minute city' was coined in 2015 by Carlos Moreno, a business professor at the Sorbonne. The idea is to design cities so that citizens have all of their basic needs, including their workplaces, within a fifteen-minute walk or bike ride from their home. Anne Hidalgo, former mayor of Paris and member of the Socialist Party, campaigned (and won) in 2020 using the concept of the fifteen-minute city as a slogan, vowing to create 'a bike lane in every street' by 2024, and to do so in large part by removing 60,000 of the city's on-street parking spots. Downtown, bicycles are now a more popular mode of transport than cars. What was happening in Paris was one part of a broader embrace of sustainable-traffic schemes during the pandemic. In France alone, 500 kilometres of new bike lanes were added in urban and suburban areas, leading to the neologism *coronapiste*, while hundreds of kilometres more were created in cities from Bogotá to Brussels, Milan to Montréal.[3] According to one study, from 2019 to 2020, cycling increased nearly 40 per cent across European cities.[4]

In Britain, low-traffic neighbourhoods, hereafter LTNs, were widely expanded. In 2020, swept along with the green zeitgeist, the Boris Johnson–led Tory government announced a £250 million travel fund for local authorities, with the explicit purpose of prioritising cycling, walking and public transport, at the same time closing roads and reducing speed limits for motor traffic. LTNs were first introduced in Hackney in the 1970s and again promoted under the Tory–Lib Dem coalition in 2014. They were generally found to be effective, with one study suggesting that car traffic decreased by an average of 45 per cent inside LTNs while increasing by an average of 4.5 per cent on roads nearby. Notwithstanding criticisms suggesting that the effect is for traffic to move from whiter, more affluent neighbourhoods towards poorer, more racially diverse ones, these schemes tended to be well supported. In the 2020s, however, researchers found that local governments were met with a 'public and political backlash', but since polls continued to show broad public support for the schemes, the exact nature of this opposition was 'hard to quantify or qualify'.[5] Such observations captured the early stages of a reactionary mobilisation that would gradually erode an emerging consensus around sustainable-travel schemes and crystallise in a series of highly visible flashpoints: first Oxford, then London.

In November 2022, Oxford City Council approved a traffic filtering system aimed at reducing car use in the city. While the proposal was labelled an LTN, the scheme itself became associated with the concept of the fifteen-minute city. The bollards and planters used to restrict car access and pedestrianise certain areas were met with a wave of vandalism. Bollard bashing spread across the country, with a single bollard in the east of Oxford – 'the most hated bollard in Britain' – repeatedly targeted: some drove over it, some managed to remove it and others even set it on fire. Leading figures in the country's anti-vax and anti-lockdown movement descended on Oxford – including stalwart of climate denial and brother of Jeremy, Piers Corbyn, and pop duo Right Said Fred – and handed out flyers mixing climate denial tropes with warnings about carbon rationing: residents were urged to 'not be guinea pigs'. A series of demonstrations reached their crescendo in February 2023, when thousands converged on Broad Street to oppose the council. As well as explicitly conspiratorial factions, neo-Nazi groups like Patriotic Alternative were also in attendance. Since the pandemic, former GB News reporter Laurence Fox explained, 'The government and the powers have gotten this desire to control our movement, our speech, our everything ... none of this is about climate.' Others in attendance decried the 'anti-motorist coalition' who were 'locking people into zones'. The increasingly fierce opposition ultimately forced Oxford City Council to reverse their plans. Elsewhere around the UK, when local councils contemplated analogous schemes, they found their meetings stormed by groups of 'freedom fighters', leading to municipalities desisting from such considerations for fear of the backlash.

That same year of 2023, the expansion of the ultra-low-emission zone, or ULEZ for short, from the inner city to greater London became a heavily contested event, quite unlike the partial expansion in 2021. As in Oxford, the anti-ULEZ campaign boasted a visible street movement, which fed out of and incorporated several local campaigns opposing LTNs around the capital. Following Mayor Sadiq Khan's announcement to expand the zone to greater London in 2022, the movement grew even more coordinated in its strategy.

Small groups regularly staged roadside demonstrations around the capital, identifiable by their yellow boards, mirroring the colour of the Gilets jaunes' disruptive opposition to fuel tax increases in 2018 and displaying messages such as 'Stop the toxic lie', 'No to ULEZ, no to pay per mile' and 'Stop the Khanage'. As preparations for the expansion commenced during 2023, a more militant wing emerged, with a group styling themselves as 'Blade Runners' vandalising the cameras monitoring vehicle compliance with their secateurs. In contrast to Harrison Ford's Blade Runners enforcing state and corporate control, London's Blade Runners viewed themselves as the resistance against it.

At the discursive level, fifteen-minute cities, ULEZs and LTNs all ended up combining with the pre-existing elements of what Naomi Klein called 'the great reset conspiracy smoothie'. Jordan Peterson, among the foremost salesmen of this noxious cocktail, sent the fifteen-minute city concept viral with a 31 December 2022 retweet lambasting the 'idiot tyrannical bureaucrats' behind the idea and advising his followers to 'make no mistake, it's part of a well-documented plan'. Leaving no doubt as to who and what is *really* behind this ominous-sounding 'plan', Peterson's post included a retweet, clarifying how 'it's already happening … #GreatReset #JailSchwab', including photos of Canterbury and Oxford's new driving zones, alongside an infographic explaining that 'the fifteen-minute city is a UN and WEF [World Economic Forum] plan because they ~~care about you~~ want you to drive less'. The tweet accrued over 7.5 million views.

Travelling on the international market of misinformation, the climate lockdown conspiracy theory spread to Canada. In February 2023, in Edmonton, Alberta – the capital of the petro province – an anti-fifteen-minute city rally opened with the Lord's Prayer. At one point, a co-organiser grabbed a large book from her backpack and presented it to a city planner in attendance: 'Would you swear on the

Bible that we're not gonna be fined, that we're not gonna have to show our IDs at stop points, or have our license plates scanned?' On the speaker's platform in front of the crowd, the bureaucrat began his oath, 'I, Sean Bohle, Senior Planner at the City of Edmonton, do solemnly swear that if the city of Edmonton ever tries to institute controls on the people's movement through the city, I will vehemently oppose it.' The crowd erupted in cheers.

The planner was being urged on by anti-lockdown activist Christopher 'Chris Sky' Saccocia. Prior to 2020, Saccoccia, the then thirty-four-year-old son of a wealthy housing developer outside Toronto, had seemingly no political experience other than a penchant for far-right shitposting. As documented by the Canadian Anti-Hate Network, he posted that 'Blacks lack the sophistication to make an advanced civilization'; that 'Muslims and child rape go together like burgers and fries'; and that *Mein Kampf* was '100% bang on' about Jewish efforts to control the world, like Hitler 'had a crystal ball into the future'. When the pandemic hit, Saccoccia channelled this world view into anti-vax and anti-lockdown activism. Growing a large online following, racking up several criminal charges and thousands in fines for breaking public health rules, he became one of the country's most recognisable 'freedom' activists, even gaining some international notoriety as an occasional guest on InfoWars, the flagship show of conspiracy theory guru Alex Jones.

In his speech at the Edmonton rally, Saccoccia retooled his pandemic-era messaging for the new climate front:

> They used COVID to get you used to the idea of being no more than five kilometers from your home ... Now, they know a pandemic can't last forever, but what can? A climate crisis can! A climate crisis lasts for generations. So COVID was a vehicle to retrain you, to set you up to accept the new climate change-enforced agenda of total control.

What this ultimately entails, he concluded, is 'the removal of private car ownership'. Sky's pandemic-era messaging, influence and following were thus repurposed for the war on the motorist. Looking to Oxford for inspiration, Saccoccia worked the crowd into a frenzy: 'You know what they're doing right now? They're smashing through the bollards with their cars. They're ripping out the cameras … And that's what we're gonna do here.' All this resistance despite there being no actual plan to introduce any such scheme in Edmonton. The Edmonton rally can be understood as a pre-emptive strike against urbanist mitigation efforts by drumming up paranoid controversies, generating moral panics and keeping constant guard against even hallucinatory signs of policies that curtail private automobility.

In this dance between the 'bottom-up' street movement and 'top-down' culture war, far-right influencers team up with long-standing denialist groups to spread the paranoia. Here it is important to underscore not just that the 'freedom' movement pivoted to climate, but that it was suffused by the climate countermovement from day one. As reported by *DeSmog*, Not Our Future – one of the leading anti-fifteen-minute-city groups in Oxford – has institutional, personal and financial linkages to organised climate denialists, including individuals connected to the Global Warming Policy Foundation, the leading denialist group in the UK. Meanwhile, such groups enjoy increased attention and traffic thanks to the likes of Jordan Peterson, who has become, in the words of climate scientist Michael Mann, a 'central cog in the denial machine', mainly by platforming and sharing such sources on X/Twitter and YouTube. And in Canada, the long-standing denialist organisation Friends of Science has been spreading the same conspiracies. Spokesperson for the group Michelle Stirling appeared on Rebel News and other far-right platforms to explain how the pandemic 'conditioned the public for fifteen-minute living', warning that vaccine passports will soon be converted into a technology to administer population-wide carbon rations.

While fossil capital has typically exercised power through more mundane tactics like lobbying, regulatory capture and funding denialist think tanks and media, its backers are now either finding or creating new forms of leverage. We are witnessing the rise of a vanguard located not in the boardrooms but online and on the streets, willing and able to serve as the 'tip of the spear' – or secateurs, as it were – for the defence of the fossilised nation.

Whose streets?

On the eve of the expansion of the low-emission zone to greater London, a few hundred protesters arrived at Downing Street with 'NO 2 ULEZ' on their mock registration plates. The Blade Runners, for their part, marked the occasion by disabling scores of cameras in what some described as the 'Night of the Long Knives', a reference to their distinctive secateurs that would seem entirely ignorant of the historical connotations. They were cheered on by far-right influencers, Tory MPs and conservative media alike. A Google map of all of London's ULEZ cameras circulated on far-right Telegram channels, the red icons indicating functional cameras, the black those that have been clipped or otherwise made dysfunctional. 'South London' – covered in black – 'looking good,' wrote the neofascist street brawler turned social media influencer Tommy Robinson. 'I'm happy for them to do it,' approved former Tory leader Iain Duncan Smith in reference to the Blade Runners' feats. Meanwhile, the demonstration outside Downing Street, despite the unremarkable turnout, was amplified on the front cover of the *Telegraph* and the *Daily Mail*. Undeterred by polls showing that the majority of Londoners support ULEZ (even those living in Greater London), this mediatised feedback loop served to legitimise, subsidise and generally encourage the growth of a more decidedly far-right grass-roots movement. The movement, in turn, created a window of opportunity for anti-ecological policies at the national level.

At the end of a hot summer of protest and sabotage in 2023, the street-level movement was given a nod and a wink by senior Tory politicians at the annual party conference. Transport Secretary Mark Harper used the stage to call time on fifteen-minute cities in which 'councils can decide how often you go to the shops, ration who uses the roads and when, and police it all with CCTV'. Having just postponed the 2030 ban on petrol and diesel cars, Rishi Sunak now announced he would be 'slamming the brakes on anti-car measures across England'. The government released its 'plan for drivers' with the objective of curtailing the ability of local authorities to either reduce speed limits, penalise driving and parking offences, implement LTNs or reserve lanes exclusively for buses. Another policy was proposed to restrict councils from using 'so-called fifteen-minute cities to police people's lives'. Then, on the final day of the conference, Sunak announced the cancellation of the northern chunk of the high-speed rail network, set to connect Birmingham to Manchester. But what to do with the freed-up funds? Of the £36 billion, nearly a quarter was reallocated for road maintenance and expansion. This was the 'right decision to make life easier for hardworking families', said the prime minister. 'We're on the side of drivers,' assured Harper. In a further mockery of the levelling-up agenda, it was announced that £250 million of the funds would be spent on filling in potholes on London's roads.

It is hard to take seriously the idea that the Tories believe low-emission zones or low-traffic schemes (which were recently and readily implemented under their watch) represent early signs of a future dystopia of 'climate lockdowns'. Using the car as a wedge, theirs is clearly a strategy designed to polarise the urban–suburban divide. For the Blade Runners and street protesters, however, there is little doubt about an impending WEF and UN-led global surveillance state in which all activity and consumption is monitored and restricted according to emissions criteria. The political trajectory and 'programme' that such a world view implies is clearly more radical than that of most Tories, yet the tail appears to be wagging the dog as the 'respectable' right loses ground, having embraced 'rougher' forces which it cannot control.

As a union between state and street was forged in defence of the car, climate protesters who opposed its free reign were met with massive repression. Recalling the moral panics and law-and-order crackdown on mugging examined in *Policing the Crisis*, a wave of bills eradicated the right to assembly and extended powers to the police, particularly in their treatment of environmental protesters.[6] Making way for the motorist, the 'wilful obstruction of the highway' became an imprisonable offence in 2022; the same went for 'locking on' to another person or object. At a summer party, Sunak thanked one fossil-affiliated think tank for their part in drafting the legislation. Climate activists would soon feel the full force of these laws. Some received multi-year sentences for climbing on bridges to disturb traffic flows, others for simply planning to do so. These were the longest sentences handed down for non-violent protest in modern Britain, prompting a UN spokesperson to call it a 'dark day for peaceful environmental protest'. In early 2026, the days got darker after the Court of Appeal 'substantially eroded' the independence of juries, following a case where five women who vandalised JP Morgan's London offices in protest of the banks' fossil fuel financing were acquitted by jury.[7] The court determined the jurors to have been improperly guided by their conscience.

Part of a broader trend of policing the climate crisis rather than mitigating it, the UK government's crackdown on climate activists was echoed by states across Europe and beyond. In Germany, the climate activist group Letzte Generation (Last Generation), following a short but intense wave of road blockades, faced street arrests, anticipatory house searches, and surveillance of their emails, chats and phone calls. Already subject to intense police brutality, the French climate organisation Les soulèvements de la terre (Uprisings of the Earth) was ordered to dissolve itself until the courts declared the move unconstitutional. Meanwhile, the French state apprehended Italian climate activists on 'anti-terror' laws at the border. In Spain, the public prosecutor declared Extinction Rebellion and Futuro Vegetal 'terrorist' groups. Meanwhile, in the US, sixty-one members of the Stop Cop City movement (also known as Defend the Atlanta Forest) had been indicted on RICO charges – a law normally used against criminal mafia organisations – to ensure the bulldozing of the forest so that a $90-million militarised police training facility could proceed as planned.

By framing climate activists as alarmists and terrorists threatening the social cohesion of the nation, an underlying alliance is forged between the far and centre right. While the inverted crisis suggests that a totalitarian state is preparing to clamp down on fossil-fuelled automobility and personal freedoms, the state is in fact increasingly crushing civil liberties in defence of fossil capital. The moral panic about the war on the motorist flips once again into a law-and-order campaign of authoritarian control. At the same time, the violence of the enraged motorist against the climate activist mirrors that of the state; in many ways, the former is deputised by the latter. Climate activists thus face a pincer-like combination of hostility and repression from both above and below.

The fossilised family

To announce his 'plan for drivers', Rishi Sunak thought it expedient to pose in the front seat of Margaret Thatcher's old Rover. 'Talking about freedom', Sunak captioned the photograph, before outlining his concerns to the *Telegraph*. The prime minister was 'particularly worried' that planters and bollards, like those in Oxford, would allow 'no vehicle wider than a bicycle to travel through', obstructing cars for families going about their lives. He knew, of course, just 'how important cars are for families'. If, following M. E. O'Brien, the family is 'one site in the broader reproduction of capitalism', then what of the family car?[8]

The Wages for Housework and Wages Due Lesbians campaigns of the 1970s offer some insight into the question. In their analysis of the sexual division of labour, 'The woman supports the man to work harder, to buy a bigger house, a car, etc., and to subjugate her needs to these needs, which are capital's.' As the capitalist system must produce cars, it must also produce 'women' and 'wives' disciplined into supporting male breadwinners, to produce 'families' capable of absorbing through consumption the over-production of materially intensive, ecologically destructive commodities and the attendant ways of life.[9]

The male-breadwinner model, however, collapsed under the combined pressures of feminist and queer liberation movements and the weight of neoliberalism, with working-class households no longer able to afford an un-waged housewife. This collapse, as O'Brien notes, 'has been a huge improvement in human freedom', particularly for women and queer people, who found a degree of independence from the personal domination of the father and husband, even as this was replaced with the abstract domination of capital.[10] The coterminous experience of increased economic precarity and radically altered gender and sexual relations, however, forms the charged atmosphere in which the far-right's revanchist call for a revitalisation of family values resonates.[11]

The dethroned entitlements of white masculinity embedded in the overturned Fordist family easily pair up with those of the driver. Such articulations gain traction from concrete historical and geographical determinations. 'The internal combustion engine', Matthew T. Huber notes, 'was the machine that guaranteed the mass dispersal of low-density, single-family housing.'[12] In other words, the geographies of atomised social reproduction are precisely those most car-dependent; the privatisation of care through the family form simultaneously entails privatised mobility.[13] To the reactionary mind, then, the urban provides a double foil for the rural and suburban: a space of relative freedom for women and queers (and therefore an anti-family space), and simultaneously one increasingly hostile to the car. Restoring the family presupposes a restoration of the geographies of familism.

It is perhaps unsurprising, then, to find familism, natalism, and climate denial intertwined in the Twitter poems of Jordan Peterson. In one, he ventriloquises the 'global fear-mongering elites' who speak from on high:

> Starve in the dark
> Peasants
> You and your swarming hordes
> of carbon-emitting children
> Cluttering up the planet.

The WEF and the UN, whom he tags, are framed as genocidal maniacs so consumed by emission cuts they would commit mass murder. The coinage 'carbon-emitting children' constructs a crisis for the fossilised family: 'they' do not want us to have children; 'they' are killing us by taking away our fossil fuels. Crucially, this appeal to 'care' for children reverses the Fridays for Future movement's central discourse, in which rapid decarbonisation is presented as a moral obligation to protect future generations. Here, the language of care is redeployed to defend fossil-fuelled family life and to frame climate mitigation as an assault on reproduction.

A similar construction is spread by David Parker, founder of Take Back Alberta, one of the most influential far-right organisations in Canada. 'Most women my age', he says,

> think the worst thing that could happen to them is to get pregnant. Okay, their careers are more important. More important than the continuation of the human race … We are living in an anti-human society that literally teaches our children that they are a disease on this planet. That the best thing we can do is depopulate. *You are the carbon they are trying to reduce*.

Through this identification of 'the people' with carbon, anti-environmentalism may lead to ultra-nationalism, and vice versa: 'You will not replace us' becomes synonymous with 'you will not replace fossil fuels'.

Cara Daggett describes a 'catastrophic convergence' between climate change, a threatened fossil fuel system, and fragile Western hypermasculinity. 'Challenges to fossil-fuelled systems, and more broadly to fossil-soaked lifestyles', she explains, 'become interpreted as challenges to white patriarchal rule.' There is a defensive kernel here – a parallel between challenges to the family unit, automobility and traditional male authority.[14]

This reactionary style sometimes turns offensive. One of Daggett's examples of 'petro-masculinity' centrally involves automobiles: in the trend known as 'rolling coal', men retool their diesel trucks to emit excess fumes, the black clouds aimed at drivers of hybrids, women pedestrians, cyclists and progressive protesters.

Or consider Andrew Tate. On 27 December 2022, one of the world's most famous misogynists tweeted at the world's most famous environmentalist, 'Hello @GretaThunberg. I have 33 cars. My Bugatti has a w16 8.0L quad turbo. My TWO Ferrari 812 Competizione have 6.5L v12s. This is just the start. Please provide your email address so I can send a complete list of my car collection and their respective enormous emissions.' The iconic response, which became the fourth-most-popular tweet of all time: 'yes, please do enlighten me. email me at smalldickenergy@getalife.com.'

Scrolling through Tate's social media empire reveals the centrality of automobiles to his persona. On the home page of Top G, his lifestyle brand selling everything from shirts to supplements, Tate shadowboxes before his mansion's parking lot filled with cars. There he is again, leaning against a black sports car on the home page of The Real World, his 'education' platform. His Rumble-exclusive show, *Tate Confidential*, revolves around luxury sports cars. The short, roughly fifteen-minute, episodes feature road trips, expeditions for new acquisitions, and Tate participating in car rallies and races. The episode titles give a sense of the sort of fare on offer, while incidentally making suitable material for a sort of found poetry of ecological sadism:

Andrew Tate tries to buy a tiger
buying every supercar in the world
the most expensive Lamborghini
buying 3 new Ferrari's with gold coins
eating camel in dubai
smoking on a plane
endless shopping spree
kicking a tree in half
eating bat meat in Stockholm
coronavirus party
setting everything on fire
we can't keep doing this.

Tate's libidinal investment in the forces of destruction is almost too on the nose. The kind of aggressive masculinity displayed is obnoxiously and transgressively nihilistic. For disaffected young men, Tate's feed is a window onto a life of unlimited hedonism that exists beyond society's dull compulsions. Living off passive income and seemingly doing whatever the hell he wants serves as a counterweight to the thwarted life aspirations of most. His refusal of social limits is matched with a refusal of ecological limits. Through an exaggerated representation of petro-masculinity, he signifies the reversal of the dethroned entitlement of traditional male authority: an attempted re-enthronement evident in his self-description as 'the *king* of toxic masculinity'. While clearly at the extreme, Tate's brand of performative excess moves beyond the defensive to defiantly affirm all that is threatened: cars, masculinity, unbridled autonomy.

There is a non-coincidental parallel, then, between 'leave our kids alone' – a main slogan of the anti-LGBTQ backlash – and a Jordan Peterson tweet demanding elites 'leave our cars alone'. Such moral panic intersections could be heard on the 2024 US presidential campaign trail, with Trump declaring that Democrats want to 'legalize drugs, shoplifting, and sexual mutilation of your children. But they want to allow your gas-powered suburban Silverados and Ford F-150s to die. Under a Trump administration, gasoline engines will be allowed, and sex changes for children will be banned,' a maxim that drew one of the loudest cheers of the night. 'Change gears, not genders,' reads one meme-turned bumper sticker. In such confluences, the anti-trans backlash is generalised to the social totality: from gender transitions to energy transitions, the point – inverting Marx's thesis on Feuerbach – is to stop it.

Zones of resentment

In the 2024 general election, Nigel Farage's Reform UK won their first parliamentary seats. Two of the five seats they won were in Essex, a county to the north-east of London, where cars play a mythic role in the region's identity.[15] Essex Man had been an archetype for the perfect Thatcherite subject: a 'young industrious, mildly brutish and culturally barren' man equipped with a satellite dish on his newly bought council house and a new car on his drive.[16] One of the MPs newly elected from this region, James McMurdock, on the buffer of the ULEZ zone in South Basildon and East Thurrock, reached straight for an auto metaphor to describe his political journey, telling the *Independent* how it felt like he had 'been on a long drive in the dark and the petrol light is flashing'. Data following the 2024 election pointed out that four out of five of the seats picked up by Reform were in coastal areas in the east of the country, which happened to have constituencies with the highest percentage of people dependent on cars to get to work.

A partial explanation for this pattern emerges from the shifting realities on the outer fringes of London and wider suburban landscapes. Since the financial crisis, car-dependent suburbs of England have been where loneliness soars, food banks proliferate and energy bills are paid late.[17] Through rising fuel prices and rental payments, the car on the drive might be less a trophy than an albatross. When it remains the only means to commute, as Giulio Mattioli and colleagues observe, 'forced car dependency' often goes hand in hand 'transport poverty'.[18] Reversing twentieth-century trends, poverty has risen faster in outer London than in its inner boroughs. While city centres are increasingly for the young and successful, and the countryside for the affluent as they age, those in the suburbs are the ones stuck in between. As Danny Dorling notes, 'The most successful hop over the suburbs when making their jump from Notting

Hill mews to north Oxfordshire cottage, from dinner parties to country suppers.'[19] Contrary to the narrative of the 'white working class' of the post-industrial North, it was suburban voters in the south of England who helped push Brexit over the line. Once the domain of prosperity and respectability, the suburbs have become places of 'political and social retreat' where people 'hark to being great again'.[20]

What factors have driven this spatial transformation? Under Fordist models of industry, where profits rested on the ability to retain earnings from the sale of commodities, land for factories and warehouses was generally cheaper. The state also applied downward pressure on the cost of land through nationalised industry and public housing. The economic and political restructuring of the 1970s and 1980s, characterised by industries uprooting and states retreating from markets, returned land as a scarce and inflationary resource. With financial returns increasingly outweighing profits from commodity production, land itself became a valuable asset and source of speculative wealth.[21] For many years, asset price inflation sustained rising prosperity, around which new electoral constituencies coalesced, bound by a shared interest in rising property prices. As economies have since been unevenly mired in stagnation, austerity and inequality, property prices have in some places stalled or subsided, a shift shown to be a determinant of populist right voting behaviour.[22] Political coalitions are coming undone at these fringe geographies, increasingly the staging ground for a resentful nationalism.

As a growing body of empirical and theoretical work suggests, it is the perception of threatened social advantage, rather than direct economic hardship, that drives much far-right support.[23] The experience of decline, then, should be understood not as an absolute measure but as a status-laden comparison. The availability of cheap consumer goods coupled with the explosion of consumer credit has eroded traditional markers of consumptive distinction, meaning middle-class households 'see their consumption behavior approximating that of poorer households' while diverging from those above.[24] As Richard Seymour writes, the emotional charge of the downwardly mobile middle classes is resentment arising from the 'fear of becoming indistinguishable from those at the bottom'.[25] Think of the quiet bitterness in seeing a gleaming SUV on an undeserving neighbour's driveway during a cost-of-living crisis. But if one measures one's fortunes against those next door, the suburbs may also define themselves in relation to more affluent urban enclaves, where unattainable asset wealth leaves many with the feeling of being locked out.[26] Social landscapes of relative deprivation are where the politics of resentment take root.

To understand how such affects cohere so strongly around automobility, we might draw on Wendy Brown's repurposing of the Nietzschean concept of *ressentiment*, which captures the mix of bitterness, frustration and resignation that leads to a wholesale rejection of prevailing value systems. The liberal, cosmopolitan values associated with urban cores – progressive gender politics, multiculturalism, green elitism – become things to be repudiated in their entirety. Low-emissions zones and low-traffic neighbourhoods are absorbed into this structure of feeling. Automobility, in this context, is felt to be a traditional right, an embattled entitlement that needs to be defended. The comparative dimension of experientially threatened status is why ressentiment so often lends itself to producerism and pits itself against parasitism.[27] Hard-working motorists are thus pitted against bike lanes, bollards and bureaucrats. As the political vehicle leading the defence, it is in these relational geographies that far-right appeals gain traction.

Britain is hardly the only place riven by such spatial antagonisms. New York City's congestion pricing system, introduced in January 2025, has triggered similar dynamics. Suburban drivers in Long Island and New Jersey, and those in outer boroughs like Staten Island and south Brooklyn, led the opposition to the first-of-its-kind programme in the US, which promises to raise billions for struggling public transit projects. In the 2025 mayoral election, Andrew Cuomo, who had supported and then flipped to denounce congestion pricing, won majorities in precincts where most people commute by car and with higher concentrations of homeowners. Zohran Mamdani, on the other hand, won the precincts where most people used public transport. Unsurprisingly, Trump waded into his hometown's war on cars. While the Blade Runners were only able to take down dozens of cameras at a time, Trump sought to outdo them by revoking federal approval for the entire system, though his intervention was then checked by the courts.

In France, the Rassemblement national have translated the low-emission zone acronym to 'zones of high exclusion'. The hostility to anti-car schemes grew to such a degree that MPs there voted to scrap them entirely, thereby 'freeing the French from stifling, punitive ecology'. One of the clearest expressions of this pattern, however, was the Berlin state election in 2023. The left–green coalition championed the expansion of public transit, cycling infrastructure and pedestrian access while at the same time making efforts to significantly reduce motorised traffic – with the Greens even advocating for a car-free city, reduction of parking spaces and a toll to enter the city centre. The CDU, on the other hand, pitched themselves as the last defence of the motorist. Warning that the red–red–green coalition intended to 're-educate Berliners as cyclists', a core conservative campaign message was that 'Berlin is for everyone, including car drivers,' with their leader posing as the latter's 'patron saint'. Incorporating messaging previously central only to the AfD's far-right programme, the CDU won the election. The map of the electoral results was striking. In the city centre, there is a smattering of red and green encircled by an almost uniform black ring, the CDU winning almost 90 per cent of votes on the city's periphery. As geographer Phil Neel predicts, where in previous periods the city centre and urban slum may have been among the most volatile sites of political unrest, in the coming decades it will be the suburbs.[28]

Some of these disaffections from below have been augmented from above. Following an unexpected victory in a by-election in the London constituency of Uxbridge, where the issue of ULEZ expansion featured prominently in the campaign, the Conservative Party came to vocally denounce the expansion. Many understood this pivot to be a political calculation: the Tories, behind in the polls with a general election anticipated, viewed this as an opportunity to signal a willingness to undermine climate policy. Party strategists had created thirty-three Facebook groups targeting different regions on the peripheries of London, each opposing the encroachment of ULEZ into their district: 'Bromley says NO to ULEZ,' 'Hillingdon says NO to ULEZ' and so on. These pages became hotbeds of conspiratorial speculation, support for the Blade Runners and racialised abuse targeted at Sadiq Khan.[29] Thousands of posts coupled their racist outbursts with anti-ULEZ sentiments: 'Fuck Ulez. If Khan gets in the PURGE must begin with NO REGRETS.' If the city is the location of green-tinged elitism, it is also the site of multicultural decay.

Roundabout

It was a Saturday in Ipswich in 2023, and 30,000 football fans were heading to Portman Road. On their way, many passed a dozen demonstrators marooned on a roundabout, waving the yellow boards typical at pro-car 'freedom' demonstrations from Cowley to Croydon. But unlike the localities where the cadres of the movement typically converge, Ipswich is not a town where any policies perceived as anti-car had been mooted. No low-traffic neighbourhoods or low-emission zones, no lowered speed limits or bike-only streets. What it did have, however, was a Tory MP declaring on television that the town 'felt like a foreign country' and claiming that development was behind the rise in high street theft. The Tory politician's intervention may best explain why this demonstration was taking place in this location: it was right across the road from accommodation used to house asylum seekers, one of the hyperpoliticised hotels where tens of thousands of refugees across Britain were placed to wait in limbo. The roundabout signals a broader shift occurring across Britain: the loose coalition of pro-car 'freedom' activists has circled back to anti-migrant politics.

Nativism had been woven into the pro-car mobilisations in Britain from the start. In Oxford, one heard how migrants would become armed enforcers of the fifteen-minute cities, locking citizens into controlled spaces. One welder of such viewpoints is the card-carrying Blade Runner Matt Hardy, who livestreams his antics to a sizable TikTok following. Based on the perimeter of London, Hardy not only delights in discovering smashed cameras, but also confronts those sent to repair them, obstructing their work and berating them for their role in supporting the supposed tyranny. He reserves particular venom for non-white workers, brandishing them as 'dinghy riders', feeding into the suspicion that migrants are in fact agents smuggled in to enact the demonic will of the globalists.

The connections between nativism, Islamophobia and cars crystallised and mainstreamed in the London mayoral election of 2024. The Conservative candidate, Susan Hall, campaigned on the promise to get rid of ULEZ and restore law and order to London. Throughout, she fixated on the idea that Labour incumbent Sadiq Khan would be introducing a pay-per-mile scheme to hurt drivers even further, despite no evidence of this being the case.

The Tory candidate also flirted with the idea that London was in reality 'Londonistan' – invoking a term popularised by a range of anti-Muslim agitators from Melanie Phillips and Douglas Murray to Tommy Robinson – in reference to the mayor's Pakistani heritage, while also blaming black communities for the city's crime.

In the run-up to the London election, the Tories published a campaign video which, rather than amplifying the credentials of their candidate, presented the city in an eerie dystopian light. Black-and-white images of deserted streets suggested that Sadiq Khan had plunged the city into darkness over his eight-year reign. A narrator with a doom-laden American accent announced how communities were being terrorised by 'squads of ULEZ enforcers dressed in black, faces covered with masks'. These shadowy state agents were at the 'beck and call of their Labour mayor master', whose 'tax on driving' was forcing people to stay inside. If anyone who had been to London recently failed to recognise anything remotely familiar about the city in the video, it is no fault of theirs. With the disregard of a high-school film project, the Conservative Party attempted to pass actual footage of a stampede that occurred at a New York subway station as evidence of London's demise. Despite the grippingly apocalyptic show, Susan Hall lost and Sadiq Khan was re-elected, foreshadowing the Conservative Party's electoral defeat in the general election and their subsequent eclipse by the meteoric rise of Nigel Farage's Reform UK.

Since 2020, Farage has positioned himself against cycle lanes and road closures. But with the war on the motorist heating up in the post-pandemic era, this became a central issue. In the run-up to the local elections in the spring of 2025, the party chairman told the *Telegraph* how Reform views LTNs 'with the same suspicion as mass immigration and net zero' and pledged to scrap them. The seats they won, however, had no LTNs to begin with, mirroring the pattern whereby anti-immigrant parties tend to receive the most votes in places with the fewest migrants. But anxiety about restrictions on car use is just one in a wider climate of decline. Another swirled around the state of Britain's transport infrastructure.

To launch their local election campaign – occurring up and down the country but not in London – Reform curated a decrepit provincial high street, a mock set constructed in an auditorium in Birmingham, the seized-up engine of Britain's auto industry. In an echo of Boris Johnson's Brexit theatrics in 2019, and confirming the shifting location of the nationalist zeitgeist, Farage arrived onstage riding a JCB 'pothole pro', a machine dedicated to fixing road surfaces.[30] On the electronic screens dangling above the stage, the number of potholes around the country looped round and round: Cambridgeshire: 64,915; County Durham: 65,038; Northumberland: 44,921. 'Aren't potholes just the perfect symbol of broken Britain?' Farage exclaimed.

If the pothole supplied the symbol, Farage provided a broader narrative connecting infrastructure neglect to anxieties over nation and immigration, state corruption and cultural decay. From the arena stage, Farage complained that mass migration has made 'our roads impossible' (the same traffic preventing people from attending that evening, he claimed). This 'invasion of our country' had also led to pro-Gaza independent MPs from 'inner-city constituencies' winning seats in Parliament. He went on to rail against young people indoctrinated at university to care about climate change, promised to go to war with those 'poisoning the minds' of the young and mocked councils for wasteful spending on cycle schemes and driving lessons for asylum seekers. We may note a parallel here with Stuart Hall's prescient analysis of Thatcherism, where he identified the prefigurative role of Enoch Powell's style of politics and the 'magical connections and short-circuits' Powellism established between the themes of race, immigration and nation.[31] Faragism does much the same, rewiring the circuitry to include climate denial.

While Farage's pothole populism might be a restaging of an already sedimented politics of decline, today it is transforming the party-political landscape. Reform rode the pothole filler all the way to landslide victories at the ballot box two months later, winning nearly 700 of 1,600 council seats up for grabs in the May 2025 local elections. Reform seized control away from the Tories in eight councils, including the Conservative redoubts of Kent and Staffordshire. This worse-than-expected result for the Tories is what Farage had been looking for, viewing this as the moment that proved that Reform had overtaken the traditional party of the right. To paraphrase Hall, one might be happier about the decline of the Tory Party were it not for the fact that so many of their themes have been rearticulated into a more radical anti-migration discourse by Reform. Roughly coterminous with this radical right transformation in the party system, street protests also radicalised into riots.

It began in Merseyside, a region in north-west England, when a spontaneous gathering outside an asylum hotel escalated into a full-blown riot. Police vans were overturned and torched; those inside the hotel built barricades to defend themselves from the mob. Though the chaos was largely uncoordinated, the neo-Nazi group Patriotic Alternative sought to claim credit. Then, following the gruesome attacks in Southport in the summer of 2024, in which three young girls were stabbed to death, riots erupted in thirty-five different locations. From Rotherham to Tamworth, Aldershot and Manchester, mosques and asylum hotels were targeted and windows smashed, with some attempting to burn them to the ground. Others set up roadblocks and ambushed cars driven by non-white drivers. Far-right groups also stormed government buildings in London. False information swirled, fuelling the outrage. The Southport attacker was said to be a Muslim asylum seeker, despite being a black British citizen born in Wales. One X user incited others to the street in the immediate aftermath: 'The war on our streets ends now, and it ends with us taking back control. Our streets, our rules, our laws, and our country.' One year on, a report suggested that a similar outburst was imminent: 'We are facing a long, hot summer, with a powder keg of tensions left largely unaddressed from last year that could easily ignite once again.'[32] Days later, sparks flared in Epping, a town in the south of Essex, on the end of the Central Underground line but just outside the ULEZ zone. A mob formed around a hotel used to house refugees, following accusations against a resident.

Throughout that summer of 2025, as protests outside asylum hotels were regularly staged, transport infrastructure became the backdrop for an outburst of nationalist sentiment. Operation Raise the Colours was a nationwide campaign to get the St George's and Union Jack flags flying. The national colours were strapped to street lamps, staked in the centre of roundabouts and draped on the railings of motorway flyovers as the display spread, in the words of the organisers, 'suburb by suburb and city by city'. Mini-roundabouts – slightly raised sloping circles painted white in the middle of roads – presented themselves as an obvious canvas on which to spray a red cross. The Labour government were supportive, describing flag flying as a 'very British way of expressing joy and pride' and encouraged more to be flown. Despite presenting itself as a spontaneous, grass-roots celebration of patriotism, the campaign was orchestrated by the far right. One organiser was revealed by Hope Not Hate to be Tommy Robinson's former bodyguard, now running security for the political party Britain First. The far-right midwifing of the project had predictable outcomes: flags were placed in front of kebab houses or coupled with racial abuse to passers-by. Across the campaign's social media networks, supporters and prominent activists could also be seen on trips to northern France, where they harassed asylum seekers and slashed dinghies. These acts were said to be undertaken 'for our grandfathers, for our families and above all for our children'.[33]

The road-painting and flag-waving campaign signalled the way to the massive Unite the Kingdom demonstration in central London in September 2025. Elon Musk beamed in as the guest of honour, calling for the 'dissolution of parliament' and for an end to 'massive uncontrolled migration'. 'It starts in London, but it will spread to every town and village in the country and no one will have any peace,' Musk exclaimed. 'Whether you choose violence or not, violence is coming to you. You either fight back or you die.' The rally was attended by an estimated 150,000 (though chief organiser Tommy Robinson reckoned 3 million). With attendees drawn from average aggrieved citizens, Reform voters and neo-fascist hooligans, it was the largest turnout for a far-right demo in British history.

In retrospect, the anti-lockdown movement's pivot to climate, targeting fifteen-minute cities and the like, appears to have been a molecular form of the general rightward transformation in British politics occurring over the last several years. If COVID-19 initially put migration on hold as the primary concern of far-right politics, swivelling instead to the mobility of the fossil-nationalist subject, many of the same themes would help pave the road back to xenophobic nativism – in effect, a far-right roundabout. The dynamic established during the pandemic and its climatic transformation into mobility-based conspiracism – a dance between bottom-up and top-down energies – continues to intensify in its wake.

More than just an accumulation of grievances, a moral panic cycle has radicalised and grown over time, transforming the party system itself. The Conservative Party under Sunak had shifted to the right in order to navigate these headwinds, only to be outflanked by Reform, who have since plundered Tory parliamentarians, councillors, donors and staff members, while eating the traditional party's lunch at the polls. Meanwhile Labour, not to be left behind, made anti-migrant posturing central to their program, initially proposing that migrants' valuables be taken at the border, accelerating the process of returning to country of origin, and coupling anti-asylum policies with diplomatic pressure on African states. Soaring past the establishment it long berated, Reform's rise is inextricable from its defence of fossil automobility as a dethroned entitlement of British life. The party's spatial politics of resentment channel a spite towards the urban centre, coded as elitist, feminist, queer, anti-family, green, multicultural and crime-ridden. The infrastructure of automobility functions as a cathexis point for nationalist investment. Its curtailment and degradation – whether through traffic zones or potholes – symbolises cultural decay and elite tyranny, tapping into nostalgia for when the nation was supposedly wealthier, whiter and working a bit better.

4
Metabolism of the Nation

In 2022, the use of trucks and tractors as a tactic of carbon populist protest became increasingly evident as a hallmark of the 'great driving right show'. In the early part of that year, hundreds of trucks occupied large swathes of downtown Ottawa for weeks, filling the capital city with diesel fumes and the blaring horns of big rigs. The Freedom Convoy triggered copycat convoys and blockades throughout the country. Ostensibly focused on opposition to public health measures related to COVID-19, it was nicknamed the 'carbon convoy' in part for its use of the red fuel can as a prominent symbol, and for its ties to a years-long protest cycle partially subsidised by the Canadian fossil fuel industry's efforts to foment a counter-movement.[1] Championed by Trump, Musk and the broader MAGA ecosystem, the convoy model was sent viral; from Washington to Paris, engines roared and exhausts spewed at carbon copy convoys against COVID-19 restrictions. Later that summer, Dutch farmers once again started up their tractor engines, as they had in 2019. This time, they spurred on a pan-European, multi-year wave of farmers' convoys. In France and Germany, followed by Italy, Spain and beyond, tractors blockaded roads and ports and circled government buildings, opposing efforts to lower nitrogen and carbon emissions.

The internal combustion engine provided form and content: a

carbon-intensive protest tactic in defence of carbon- and nitrogen-intensive production. A similar class composition was evident too, with each cycle containing elements of corporate subsidisation, yet finding a relatively autonomous mass base in the fossilised petty bourgeoisie: owner–operator truckers, small and medium-sized farmers, small-business owners with ties to the oil and gas industry. These well-resourced 'energetic' class fractions proved capable of sensational protest, utilising processions of large machinery as 'a tool of infrastructural occupation' in defence of high-emitting transport, energy and food systems, and their role within them.[2] The Freedom Convoy and related protests became one of the largest and most significant right-wing protests in Canadian history, leading to the government's unprecedented use of the Emergencies Act to quell the protests. The European farmers' cycle achieved what few social movements have since the 1960s: a transnational project with a truly European dimension.[3] Stalling traffic, blasting horns, dumping waste and flashing headlights, the tractors sprayed shit on the institutions of the state, both figuratively and literally, while rapidly forcing significant concessions.

The convoy thus became the mode par excellence of the climate countermovement's pivot to protest, with the fossilised petty bourgeoisie serving as something of a 'Goldilocks' class fraction for climate obstructionist purposes: economically subordinated enough to pass as the 'people', dependent enough to fiercely defend the order that sustains it, and, crucially, in possession of the machinery necessary for spectacle. To the media, hundreds of trucks or tractors appeared to suggest a popular uprising in a way that hundreds of thousands of bodies – whether for the climate or for Palestine – could not. Emanating from the convoys was not only diesel exhaust, but the 'disorderly emissions' of moral panic and ideological radicalisation, as far-right actors both stoked the movements and directed them onto nationalist terrain.[4]

While the focus of the farmers' movement was on issues pertaining to the production of food, moral crusaders linked the plight of the farmers to the pits of people's stomachs: 'No farmers, no food', ran the movement's tagline. Convoys were not only a method for defending and deepening the metabolic rift within carbon and nitrogen cycles, but also framed as a defensive campaign against globalist plots to phase out meat consumption in favour of insect-based diets, with disgust mobilised to connect the dots between invasive species and population 'replacement'. Riding pandemic-fuelled conspiracism, this dovetailed with cults of bodily purity and renewal: a corporeal anti-globalism crystallised into carbon-intensive diets promising national rejuvenation. At the more extreme end, and most explicitly with the farmers, these territorial associations blended into fascistic themes of 'blood and soil', stark gendered binaries of rugged masculinities and supportive 'trad' wives, alongside wellness obsessions orbiting around raw-food diets and extreme fitness regimens. The convoy-based protest cycles thereby accelerated a broader moral panic cycle, defending energy-intensive metabolisms framed as central to the vitality, strength and virility of nationalist bodies: the metabolism of the nation.

Rural revolt

During the 2010s, insects as foodstuffs became a staple in debates about sustainable futures. For their high levels of protein and low land use demands, particularly in relation to current methods of livestock farming, crickets were lauded as a potential meat substitute, courting media and organisational attention. The UN published a report on the subject; the pages of the *Economist* and the *Wall Street Journal* ran features. One insect-farming start-up received a six-figure donation from the Bill and Melinda Gates Foundation. Subject to the pandemic swirl of misinformation, this future food scenario became the object of spiralling scrutiny and suspicion. The main lines of interpretation were that evil elite actors intended to force insects down the innocent majority's throat under the pretence of a climate crisis. Having forced vaccines into all our bodies, food would surely be easy. The theory dovetailed with other COVID-era inverted-crisis spin-offs: insects were seen as the natural diet for a fifteen-minute prison, as customarily warned on memes and protest flyers. Eating insects, according to some commentators, had overtaken pizzagate as the online right's conspiracy theory *du jour*. But it was the farmers' discontent that allowed this conspiracy to grow deep within 'meat space'.

On an October morning in 2019, 2,000 farmers in the Netherlands decided they would not be tending to their duties. Instead, they secured flags and signs to their tractors, climbed into the cabs of their trucks and chugged down the highways from their respective farms to their country's political capital, The Hague. For cars and other road users, the urban sprawl along the country's western coastline had come to a standstill, causing the longest traffic jam in Dutch history.[5] Journalists were lured to the gridlock, their cameras pointed towards the handmade signs fixed to the incongruously sized vehicles: 'How dairy you', 'Farmer and proud' and 'No farmers, no food'. The primary source of the protesters' ire was their government's hastily enforced regulations to reduce ammonia and nitrogen oxide emissions, the two leading national sources of which were farming (110 million kg/year) and road transport (38 million kg/year), with the Netherland's emissions per square kilometre being the highest, several times the EU average. Among the policies proposed were reducing livestock numbers, cancelling an airport project and lowering speed limits on national highways – measures that sparked protests in which farmers were heavily represented. The opposition proved prefigurative. Dutch farmers' tractors not only blocked roads at home, but charted the course for rural resistance across Europe.

Towards the end of 2023, thousands of tractors converged near the Brandenburg Gate in Berlin, protesting the traffic-light coalition's proposed scrapping of subsidies on red diesel used for off-road machinery in agriculture and construction. At the turn of the new year, a flashpoint in the protests confirmed their unpredictable and confrontational nature: for hours, several hundred farmers blocked a ferry transporting the then Minister of Economy and Climate Robert Habeck as he returned from a family holiday on a North Sea island. Next, in France, farmer fury was refracted through several issues, such as cuts to diesel subsidies, low crop prices, competition from foreign foodstuffs and slow payments from the EU. Here too the protests were scattered, with tractors and hay bales used to obstruct major highways and supermarkets around the country. In the early months of 2024, tractors motored through Britain, Belgium, Cyprus, the Czech Republic, Denmark, Greece, Ireland, Italy, Lithuania, Poland, Spain and beyond. In each instance, farmers were motivated by a country-specific bundle of grievances, from cheap imports and low prices to waning fuel subsidies, yet for many involved the road ultimately led to Brussels and its perceived regulatory overreach.

What had started as a throng of tractors in the lowlands had become a truly pan-European protest wave capable of wringing considerable concessions out of the state – in the EU, most notably the withdrawal of Green Deal legislation, thus eliminating directives to limit pesticide use as well as targets to curb agricultural emissions.[6] Further rollbacks were made at the national level too. In the Netherlands, the government promised not to reduce herd sizes or close farms near conservation areas. In Germany, the farmers won a delayed phase-out on fuel subsidies. In France, plans to lower agricultural fuel subsidies were sacrificed almost immediately, while national efforts to reduce pesticide use were also halted. The farmers, as an energetic class fraction, were broadly successful in wielding their tractors and privileged position at what Elizabeth

Chatterjee calls the 'diesel–highway–food' nexus to force concessions, ensuring that policymakers would pause before considering mitigation measures in the future.[7]

The farmers and their tractors were an engine not just for rolling back policy, however, but also for pulling electoral politics to the right. Across the cycle, the far right sensed that real momentum could be gained by harvesting the political energy rising from the fields, with platforms adapted accordingly. Again, the Netherlands showed the way. As the traffic jams lengthened that same October morning in 2019, a new political party was signed into existence: the Farmers–Citizens–Movement party. Over the following days, weeks and months, as the tractor turmoil expanded, so did the party's profile. Established as the political wing of rural rage, by the spring of 2023 the party had won landslide victories in the regional elections across the country. Soon thereafter it again performed strongly in the national elections, helping to form a (short-lived) coalition government with Geert Wilders's PVV in 2024.

The farmers' struggles both mirrored and extended other post-pandemic modes of political mobilisation. In Britain, the ground for popular support for the farmers to take root had been prepared by the war on cars. Even while the grumbles of British farmers remained relatively muted, the 'No farmers, no food' slogan had already found its way onto flyers for anti-ULEZ street demos. In addition to star appearances by popular Blade Runners, attendees at these jamboree-style events could also find tractors on the suburban edge of London. As protests escalated following Starmer's first budget, one organisation, Together, was on hand to coordinate the political energies: building stages, inviting speakers and printing placards. Formed in the wake of the pandemic lockdowns to challenge the imposition of vaccine passports, the group was instrumental in the fifteen-minute city and ULEZ spectacles while orchestrating other online campaigns to challenge net zero targets.

In January 2025, standing on a Together stage constructed on the edge of London, Nigel Farage, decked out in tweed flat cap and Barbour jacket, addressed an audience of tractors. The government's numbers do not add up, he insisted. 'Could it be they just want lots of land? Because they're planning on letting in another 5 million people and want to fill tens of thousands of acres of agricultural land with solar panels?' Farage was not alone in his reasoning. Jeremy Clarkson, the former *Top Gear* host, now rebranded television farmer and staple of the protests, speculated in his weekly column in the *Sun* that the state's sinister plan is to 'ethnically cleanse the countryside of farmers' to make way for immigrants and wind farms. These were some of the ideas that accompanied the tractors as they rolled up on Westminster. In Poland, the connection between agriculture and migration was made more explicit after farmers allied themselves with the Civic Border Defence Movement and drove their tractors to patrol the border.

In Canada, farmers and tractors had already played a small part in the Freedom Convoy and associated protests across the cycle. Trucking for 'closed borders, open pipelines', the February 2019 United We Roll Convoy to Ottawa included a large grain trailer with a decal reading 'Support Canadian agriculture – carbon taxing destroys Canadian economy – put the grain in the train and oil in the pipe.' It also featured a little cartoon mascot: a smiling steak in a cowboy hat encouraging the viewer to 'eat beef'. The convoy had grown directly out of the oil and gas industry's effort to incubate a pro-fossil 'movement' willing and able to 'take to the streets', yet intersected with growing far-right nationalist groups, rebranded as Yellow Vests Canada.[8] The Canadian Yellow Vests proved capable of mobilising autonomously from their corporate benefactors, taking the convoy beyond mere 'astroturf' and serving as a training ground in which future freedom movement and convoy organisers would cut their teeth.[9]

Given the partial overlaps with farmers in the Canadian cycle, it was unsurprising to see transatlantic solidarity. The red maple leaf could be spotted amid upturned Dutch flags across the 2022 Netherlands farmers' protests, and the Canadians reciprocated in turn, bringing trucks, tractors and Dutch flags to Ottawa and provincial legislatures.[10] Wishing to confirm the cultural exchange, one prominent voice of the Freedom Convoy declared, 'We stand proudly with Dutch farmers in the continued fight against government overreach and the globalist elite. Welcome to the revolution.'[11] Some months after the Freedom Convoy, leading organiser James Bauder arrived at an Edmonton protest in solidarity with the Dutch farmers. 'Trudeau is already a proud supporter of the WEF and UN Agenda 2030,' he told far-right outlet Rebel News, and so a similar 'Nitrogen Green Deal' would likely be forthcoming in Canada. 'Backing farmers because I don't want to eat bugs,' read another Albertan convoyer's protest sign, alongside another that doubled down: 'All beef, no bugs'.

Producers versus parasites

Conveniently, rural discontent fits well within the mobilisational frames of the far right. The image of hardworking, traditional rural folk can be counterposed to an out-of-touch urban elite, an antagonistic rift between the countryside and the city that is ripe for exploitation. Yet in practice the protests were not primarily driven by struggling small farmers but by very different interests. It was rarely the often immigrant workers who do manual labour on the farms turning out to protest, but rather the landowners themselves.[12] In the Netherlands, large agribusiness sponsored several protests, with one meat-processing company, Vion, even handing out free hamburgers. 'Take one while you still can, if it was up to the government, we wouldn't be able to eat them anymore,' urged one employee.[13] Still, the populist imagery held. The protests often called for more food sovereignty, a theme which at once echoes a left tradition of anti-colonialism, chiefly in the global South, yet also fits neatly with the far-right's anti-globalism. In the hands of far-right agitators, therefore, disquiet in the countryside – blamed upon the state and its political representatives – can be woven into the nationalist narrative arc: the nostalgia for past splendour and the promise of future restoration.[14]

Both the class interests of big fossil capital and the ideology of the far right articulate well with the concrete 'social force' of the fossilised petty bourgeoisie.[15] While climate activists who use their bodies to block traffic are labelled terrorists, police and politicians adopted a permissive attitude towards the farmer's convoys, which faced minimal repercussions, and even tacit endorsement. In Germany, Last Generation highlighted this hypocrisy by bringing toy tractors to their roadblock protests. 'Listen to us, we have tractors!' one activist painted on their sign. Indeed, the tractors seemed symbolically charged, the moral economy of Fordist 'petro-farming' translating itself into a 'moral forcefield surrounding farmers' protests, defending them from criticism', as a frustrated George Monbiot observed.[16]

The class composition of contemporary agriculture reveals a landscape in which large multinational agribusiness companies are interwoven with highly capitalised, hyper-consolidated, carbon-intensive farmers who, far being from peasants or labourers, are 'more akin to business professionals'.[17] Notwithstanding decades of hyper-consolidation and capitalisation, however, farmers could still stand in as modest producers for the nation. It was this ideological work that the farmers did for capital which, for Marco D'Eramo, explains why the protests were received so favourably: 'This abstract financial system needs to anchor itself deep in our psyches in order to effectively govern at the level of the nation state.' The symbolic investment in farmers, their work and their produce is vital for the cohesion of national identities, naturalising the 'imagined community that is created around the potato, the grape or the white asparagus'.[18]

As the burden of ecological reform falls heaviest on small and medium farmers, unable to reap the economies of scale achieved by vast industrial landowners, a kind of agricultural petite bourgeoisie emerges as a new carbon populist subject with distinct material investments in the status quo – a position for which the far right's underdog interpellations are well pitched. Squeezed between reliance on big fossil capital (while lacking its adaptive capacity) on the one hand, and the necessity of rapid energy transition on the other, these contradictory classes are breeding grounds for economic anxiety and ripe for reactionary recruitment – offering a 'warm humus for ideology', as Ernst Bloch described such classes.[19] As they drive rightward, the fossilised petty bourgeoisie inject organised denial and delay with the 'fanaticism of the middle classes' along with extensive mobilisational resources.[20] This class fraction not only puts a humbler, blue-collar tint on economic demands primarily benefiting big capital, but also offers a whiter hue with which the far-right can mobilise around 'pure' people rooted in the heartland and assailed on multiple fronts.

Here, conspiracy theories regarding insects perform an important ideological function. 'Made in Italy vs. Insetti' was the slogan of numerous rallies organised by the Italian far right, which spread moral panics over globalist insects and lab-grown food that 'has no relationship to man, land or work'. 'Let's change Europe before it changes us,' read one Lega billboard depicting a man eating a cricket. In France, Marine Le Pen similarly warned that Europe's technocracy would soon be compelling citizens to eat insects and, inverting the trendlines on forced displacement, stated that to be exiled 'is not to be banished from your country, but to live in it and no longer recognize it'.[21] This complemented an online disinformation campaign tied to the Russian state, which stirred up a panic about bedbugs infesting France, tying them to the rise in Ukrainian refugees. From demography to diet, fear of the foreigner was translated into pest control. And in the Netherlands, farmers' protests adopted a similar entomophobia, advertising upcoming actions with crickets on forks and the slogan 'Vote them away'. European elections represented an opportunity to spray political pesticide on the traitors.

The documentary *No Farmers, No Food: Will You Eat the Bugs?*, released in 2023, furnishes the conspiracy theory with a shiny 'info-tainment' gloss. Setting out to expose 'the hidden agenda behind "Green Policies" that are pushing people to eat bugs', the film cuts from shots of politicians at global summits discussing sustainability goals, to giant saucepans sizzling with fleshy grubs, to distressed farmers apparently facing financial ruin. 'They're not shutting down the big corporate farms but the small to middle-sized family-run farms,' clarifies stalwart of climate denial Marc Morana.[22] The documentary was produced by Epoch Times, a US-based far-right media network organised by the Chinese anti-communist Falun Gong movement. The daily newspaper counts Robert F. Kennedy Jr and Stephen Bannon among its readership, while its documentary wing produces films with titles like *The Spirit of Texas*, *The Dark Origins of Communism* and *ISIS Sex Slaves*. Oscillating between the US, the Netherlands, Sri Lanka and beyond, the *No Farmers, No Food* film provides the farmers' crisis with global coordinates: a 'master plan for humanity' devised by the UN (or was it the WEF?) and enacted by treacherous governments. Thierry Baudet and other members of his party are on hand to confirm that this is exactly what is happening in the Netherlands. It's got nothing to do with the climate; instead, they want to get rid of the farmers. But why? Forerunners to Clarkson and Farage, one widely circulated theory in the Netherlands insisted that the government wants to evict farmers to make space to build apartment complexes to house refugees.[23] Through the frenzied clouds of crickets and locusts we may detect a recognisable conspiratorial form: nefarious elites introducing foreign species to undermine an innocent society. The great farmer replacement is under way.

The documentary host's over-egged revulsion at discovering that his chips contain insects suggests the role of disgust in this rhetoric. Disgust has evolved over millennia to guard against disease, anchoring this reflex in primordial depths.[24] Insects, for their association with waste and infection, are apt to elicit such a reaction. Socially, disgust helps establish 'self-protective boundaries', structuring what is desired and what is not, what is pure and what is polluted, what is native and what is foreign, what belongs and what is excluded. Psychologists thus consider disgust an 'exclusionary emotion' that facilitates 'outgroup dehumanisation' and may, when politically charged, justify the 'notion of "cleansing" an ethnic group of elements suspected of polluting its purity'.[25] It is through this elementary kind of revulsion that the human–animal divide is said to dissolve, enabling the persecution of minorities to be moralised and accepted – a logic that has historically underpinned genocidal language.[26] While the documentary in question does not reach such extremes, it trades on similar affective cues.

The nationalist right has long used similes and metaphors related to the natural world to dehumanise migrants ('*rivers* of blood', 'our country *swamped*'). Linking together aspects of different moral panics, such discourse often has insects playing a central role. When speaking about migration, tabloids and politicians invoke the imagery of 'swarms', 'infestations', 'ants' and 'cockroaches'. In the build-up to the Brexit referendum, as cultural studies scholar Jonathan Davis has argued, the tabloid press used 'highly selective coverage of invasive species' as a way of 'consolidating their anti-immigration narratives and association of Brexit with tighter border security'. Headlines warned of 'Euro Moths' and an 'Army of Turkish SUPER ANTS' out to ravish British nature; 'Vote leaf to protect our country … and our cabbages,' implored one slogan. Flora and fauna were used to reinforce 'a stark binary between the naturally foreign and the naturally native', yet the symbolism of invasive species ran deeper. Portrayed as a 'teeming mass with unified intention', the threat that insects signify corresponds to the conspiratorial fears of orchestrated demographic and cultural replacement in which migrants play a knowing role.[27] In that same campaign, Farage posed in front of a billboard with scores of refugees on the move. This was the infamous image that, as Sivamohan Valluvan observes, rendered the group 'as insects, as pests, as grotesque', the embodiment of 'lumpen decay and pestilence'.[28]

Often, more explicit associations are made. Beau Dade, regular host on the inaptly named *Lotus Eaters* podcast, was blocked from standing as a councillor for Reform UK following an exposé of an article he had penned, which claimed that Britain's economic and social problems would be ameliorated if the country 'rid itself of the foreign plague we have been diseased with'.[29] In a still uglier episode, thirty-two-year-old Callum Parslow entered a hotel known to house asylum seekers in Worcestershire in the spring of 2024. He was 'angry and frustrated' by the Channel crossings and wanted revenge, so he stabbed a man eating his lunch. He had drafted a post for X on his phone, tagging Tommy Robinson, Keir Starmer, Rishi Sunak, Nigel Farage and Suella Braverman. 'I just did my duty to England,' he wrote, adding, 'they will call me a terrorist, they will call me an extremist: I am neither. I am but a gardener tending to the great garden of England. I removed the weeds; I exterminated the harmful, invasive species.'[30]

In the generalised conspiracy of the inverted crisis, the rhetoric of purity and pollution veers towards extermination. Since 7 October and the commencement of a genocide in Gaza, utterances of a different kind of genocide have also become increasingly audible. Patrick Moore, forever dredging up his 'co-founder of Greenpeace' credentials, warned that bans on nitrogen fertiliser, agricultural inputs that over half the world's population depend on, revealed the climate movement to be 'truly a death wish in disguise, and the disguise is to save the Earth'. In Canada, prominent freedom and Albertan separatist leader Dennis Modry similarly declared the 'WEF–UN'-led agricultural transition to be 'genocidal', framing separation from Canada as a way of avoiding such a fate. (Modry reports having received assurances from Trump's State Department that the US will recognize an independent Alberta, which the department views as 'essential for energy dominance', since the province is the leading exporter of oil to the US and home to the world's third-largest oil reserves.)

But conspiracist-in-chief Alex Jones revealed the full blueprint of the globalists' wicked plan. In an interview with Tucker Carlson, Jones communicated the familiar lines of fifth-phase denial: 'We are going to have a post-industrial world by 2030, we are going to have no personal cars by 2030, you will be eating bugs by 2030.' After acknowledging that Jones had predicted the events of 9/11 days before they happened, Carlson, attempting candour but exhibiting smarm, wished to know what was going to happen next. It was then, while peering into his crystal ball, that Jones revealed how the globalists will soon 'start the depopulation' of humanity to 'bring it down to 500 million'. How will this go about? Well, the globalists are bringing in 'tens of thousands of military-aged men from the Middle East' to enact the culling. The terror is on its way. 'But I thought we were opposed to genocide?' pitched in Carlson, *faux* naively. Streamed exclusively on Musk's X, this video was watched nearly 50 million times. Already in 2019, genocide inversion was being spewed along with the exhaust of the Dutch tractors, with one representative of the Farmer's Defence Force remarking, 'If there will soon be no more farmers, don't say "*wir haben es nicht gewusst*" ["we didn't know"].'[31]

Corporeal anti-globalism

The threat of invasive elements does not only belong to a dystopian future. The import of foreign foodstuffs has been a source of contention across the continent during this cycle of rural resistance. Farmers in Poland drove their tractors to the south-eastern border to block the imports of cheap grain coming in from Ukraine. In Spain, tractors and trucks blocked logistics nodes to protest cheap oranges coming in from North Africa. 'No to N.Z. Lamb!' read one banner by the port of Dover in the UK. A lorry in France discharged a container-load of imported foods in the middle of the road, owing to its non-national origins. These statements can be interpreted in several ways, one of which – the economic justification – is within the easiest and most satisfying reach: foreign foodstuffs are undercutting our produce on price and this is unfair. But farmers and food are intimately tied up in processes of national identification, the fare of supermarket branding and prime time cooking shows, filed dutifully by scholars of nationalism under 'banal nationalism', 'everyday flagging' or even 'gastronationalism'.[32] In the hands of the far right, such acts of defiance can do important political work.

As certain foodstuffs – meat and dairy, in particular – are brought into question by climate change, national investments in food are mobilised in defiance of climate policy. 'Food sovereignty' has been a central pillar in Giorgia Meloni's nationalist project, and in late 2023 her government banned lab-grown meat, labelling it a threat to Italian culture. Former leader of the British National Party Nick Griffin, who led the post-fascist party to its best electoral performance in the European elections in 2009, now cooks up meals using strictly British ingredients and posts tutorial videos online. And in

Germany, such meat populism had already emerged as far back as 2013 in response to a Green proposal for weekly 'Veggie Days', becoming the single-most discussed issue of that year's election campaign and the cause of the party's precipitous slide in the polls. More recently, AfD leader Alice Weidel champions carnivorous common sense, wagging her index finger and shouting, 'They will not take away my schnitzel!'

The strength of attachments to national foodstuffs appears to be realised in reaction, only made explicit when perceived to be under threat. With the spectre of climate regulations, such fears feel more immediate. But even in the absence of any concrete plans or policies, speculation and fiction may do political work, as in 2023 when Rishi Sunak alerted the public to Starmer's plans for a meat tax, despite no such policy having been proposed. From the GOP to the Tories and beyond, such pre-emptive forewarnings of pending restrictions on meat consumption have been part of the right-wing repertoire for much of the last decade, tilling the soil for any eventuality. In these pre-emptive strikes against the hallucinations of a climate movement, any whiff of encroachment on total market freedom is spun into a deadly storm coming for the nation, organically bound to the land and those who work it.

To observe how national investments find their way into foodstuffs perceived to be under threat, we can consider a substance few would likely imagine to be a nationalist mobilising device: milk. During the farmers' protests in the Netherlands, as Harriet Bergmann and Oscar Talbot argue, milk emerged as a cultural signifier of 'white innocence' used to distinguish 'the rural hard workers from the urban oat milk elite'.[33] Milk was viewed as a 'common sense good thing', tied to mothering and childcare, the latter association nurtured by the free school milk scheme at elementary schools, though not all schoolkids participate. Lactase, the enzyme that breaks down cow's milk, is more developed among populations of

Northern European heritage than in other parts of the world. Part of this populism by milk, then, is 'corroborated by the suspicion that Muslims don't consume dairy products'. In Britain, Jacob Rees-Mogg, responding to reports on declining milk consumption, attempted an inverted Orwell, complaining how 'only liberals drink skimmed milk to go with their faux leather sandals.'[34] 'Full fat, creamy milk', he went on, 'will nourish your inner Tory'. Not to be outdone, Farage posted a video raging in a hotel restaurant, where only semi-skimmed and plant-based alternatives were available. 'Why have they only got bloody left-wing options!' he railed. More than glib provocations in a tired culture war, milk, as a national signifier, flows from historical depths.

Like cars, milk flourished with the nation state in its post-war development phase, with state-supported agriculture as part of a generalised push for self-sufficiency in food. So instrumental was the British state in the promotion of milk that a connection was forged between pure milk and modern, healthy citizenship.[35] Subsidies led to overproduction and surplus produce was often soaked up by kids, with free milk at school a standard in many advanced capitalist economies in the second half of the twentieth century. In comparison to today's standards, vast quantities of milk were drunk. In the UK, for example, from the mid-1950s to the mid-1970s, on average 2.7 litres of milk were drunk per person per week, compared to only half that in 2020. Over the course of these creamy decades, the idea that milk was the provider of vital protein, strong bones and general strength was solidified. In the 1980s, the Dutch dairy board embarked on an advertising campaign. Circling an M in the motif that resembled a car brand, 'Milk. The white motor', it read beneath, a slogan which has become a cultural touchstone of sorts, recalled in the contemporary to describe something that provides a boost of energy or vitality. Milk, on this reading, may speak to nostalgia for lost national strength – a post-war milky melancholia.

In the aftermath of the pandemic, associations of milk and strength have again bubbled to the surface in sometimes strange ways. To some, milk was a natural antidote to the invasive species that was the virus, a more accomplished fighter than the untested vaccines. On the grill of one truck at the Canadian Freedom Convoy were the words 'When milk is your booze, corona will lose'. In 2024, yet another conspiracy moved from the screens to kitchen sinks. In this tale, Bill Gates is forcing a feed additive called 'Bovear' upon dairy farmers, producing unsafe meat and dairy which could lead to male infertility, forcing people to boycott particular brands or pour existing supplies down the drain. Gates's malice to reduce the population was understood as part of a wider plot to control the food supply, to replace meat and dairy with fake alternatives, insects included. The truth is more depressing. Because of the notoriously high levels of methane released by cows, the Bovear additive attempts to reduce the gassiness of herds – a technological fix applied in the belief that capital and the climate can peacefully coexist.

The popularity of unpasteurised milk has risen in these same years, in tandem with a discursive suspicion of multinational corporations on the far right. This is one of the ways the far right is staking out positions typically occupied by the left, in this case an Occupy-inflected corporate distrust. One recent trend sees users on X filming themselves picking apart steaks in restaurants, showing how what poses as an expensive cut is merely a veneer concealing cheap, processed meat inside – exposing the manipulative work of Big Agro. But the ideological scrambling runs deeper. Within social movements associated with health, nature and fitness, a traditionalist rejection of modernity merges with romantic ideas of nature. 'Conspirituality', 'dark hippies', MAHA and 'granola nazis' are some of the markers offered to signal where the world of wellness blends into whiteness.[36] This ecological rebranding on some parts of the far right, then, aligns them nearer to some neo-Nazi and eco-fascist cults, who revere natural hierarchies, organic orders and bodily purity. It is at this end of the spectrum that food becomes a source of ultra-national nourishment.

Within this diagonalist milieu, criticism of the chemical business, the pharmaceutical business and large agribusiness mixes with extreme fitness and carnivorous masculinity, at once seemingly critical of capitalist food systems, yet amounting to a mythical veneration of carbon-intensive diets. Eating meat becomes understood as a way not only to own the libs or to mock the soy boys, but also to rejuvenate the nation. Jordan Peterson, for example, now only eats beef, a 'lion diet', that he claims has helped him deal with insomnia and other mental health issues. But Peterson, along with Joe Rogan, who introduced his massive audience to the diet by trying it himself for a period, represents merely the starter pack of #CarnivoreDiet influencers. For those further along the radicalisation pipeline (or food chain, as it were), one may turn to the Liver King, a huckster of supplements and liver bars who directs his followers, dubbed 'primals', to follow his 'ancestral lifestyle', racking up hundreds of millions of views for videos pushing the exclusive consumption of proteins only available to Palaeolithic man: internal organs, bone marrow and uncooked flesh and eggs.[37]

Higher still up the extremist food chain are the formerly anonymous fitness fascists Bronze Age Pervert and Raw Egg Nationalist.[38] These leading figures of the right-wing bodybuilding community promote raw milk, eggs and organs, too – but articulate this within an ultra-nationalist body politics premised on preparation for violent, white nationalist revolution.[39] The Bronze Age Pervert, or BAP, as he is commonly referred to, is a pioneering figure within this subculture. In the world view of BAP, diet is centrally linked to millenarian decline and rebirth. Mobilising Evolian theories of civilisational cycles, BAP attributes the present degenerate stage to weak men, poisoned by 'xenoestrogenic' environmental pollutants and subordinated by the 'gynocracy'.[40] 'Real men', and above all the '*white race*', require space, *Lebensraum*, but are hemmed in by a 'longhouse' culture ruled by 'matriarchs'. Restoration awaits the formation of homosocial 'gangs' – *Männerbünde* – which he describes as 'brotherhoods of savage men who have decided to purify the earth and rid it of the infestation of the human-cockroach'.[41] The increasing prevalence of fascist fitness or 'active' clubs owes much to BAP's masculinist and racialised vision of renewal, as training is seen as essential for the pending 'maximum stress situation' in which 'young men and their captains' will be called upon. In this vision, extreme diet and fitness are inextricably tied to the cultivation of a vanguard capable of carrying out a revolutionary project, restoring order by implementing 'real' (i.e., military) government. The inverted crisis is here gendered and engendered: the present state of supposed civilisational decay is attributed to effeminate men (the atomised liberal masses referred to as 'bugmen'), who must be violently subordinated to the physically fit.

Largely eschewing politics for metapolitics, Bronze Age Pervert and his leading epigones like Raw Egg Nationalist (REN) do not concern themselves with the protest cycles foregrounded in this chapter, but their masculinist gastro-civilisationism has been mainstreamed by some of the very same platforms who most vigorously championed the freedom and farmers' movements. For example, although his books are released through obscure neo-Nazi publishers like Antelope Hill, Raw Egg Nationalist regularly writes for *Epoch Times* and *Info Wars*, and was prominently featured in Tucker Carlson's 2022 documentary *The End of Men*. He features heavily in a segment of the film on 'bro scientists' dedicated to testosterone or 'T-maxxing': removing parabens, BPAs and phthalates from one's environment; eliminating dependence on industrial food systems and instead embracing regenerative farming; 'slonking' copious raw eggs and meat (vegetarianism of course 'tanks your testosterone');

and 'doing something hard every single day', such as taking cold showers or sleeping on the floor, since 'comfort kills'. The enemy of this 'ancestral' masculinity, of course, is 'soy globalism'. In REN's gastronomical variation on the great reset's canonical maxim, the globalists command the people, 'Own nothing, live in the pod, eat the soy.'

Most alarmingly, *The End of Men* is bookended by a BAP-ist meta-narrative according to which 'hard times' are caused by 'weak men', requiring 'hard men' to 'restore order'. In combination, the meta-narrative of Evolian cycles combined with REN's 'bro scientist' prescriptions for T-maxxing amounts to a palingenetic corporeal mythos in which national rebirth depends on millions of micro-fascistic alterations to the daily routines and subjectivities of white men, with a *telos* of vigilante violence. What is called for is something like a post-postmodern, irony-poisoned version of *Freikorps* subjectivity, positioned as a bulwark against the twin floods of feminism and phantasmatic communism.[42] *The End of Men* concludes by first inoculating the audience – 'Do not listen to the libs,' viewers are instructed in so many words; 'There is nothing fascist about getting fit and healthy with your buddies' – then immediately pivots back to the meta-narrative, instructing viewers that 'well-ordered, disciplined groups of men bound by friendship are dangerous', and that even 'a few hundred men can conquer an entire empire'. The mythology of civilisational cycles, then, spurs on the formation of a corporeal vanguard hardening their bodily metabolism, working up the stomach for palingenetic violence.

If politics, for the *nouvelle droite*, is the occupation of territory, and metapolitics the occupation of culture, then across these cycles of protest the two seem to intersect. Food itself may dissolve such boundaries: it is eminently territorial and cultural. Notwithstanding its many contradictory elements, the Dutch farmers' protest has been boiled down to a defence of 'the assumed right for continuous expansion', an assertion of spatial jurisdiction communicated through a spectacle only internal combustion engines can conjure.[43] Fordism, Huber reminds us, 'was not only about homes and cars', but also about certain technologies of food production and norms of consumption.[44] The fossilisation of agriculture and the concomitant moral standards of consumption – historically unprecedented quantities of meat not least – have only intensified through the neoliberal period of farm consolidation, requiring constant expansion, following the logic of capital accumulation. The contradictions between this mode of food production and the smaller farmers – including eco-fascists who fetishise the pure and organic – seem not to weaken but to *strengthen* the movement as a whole, while subordinating any genuinely ecological potentiality.[45]

Of course, by way of their constitution, the far right is unlikely to challenge the capitalist social relations that coordinate the robbery of the soil. In the far-right's fantasies of a nationally mediated metabolism, always absent are the migrant labourers who work the land. Always ignored is the tumult that climate disaster brings to fragile and globally integrated patterns of production and supply. Rarely discussed are the disciplining compulsions determined by a global food system, and the centrality of cheap food for social reproduction. If present trends are anything to go by, the far right will instead continue to drive the inverted crisis into power, linking all grievances to a totalising globalist agenda, while defending the drivers of social hardship and ecological degradation. The attachments and values championed by the far right are realised only in reaction, mobilised in defence of the fossil-fuelled status quo when it is brought into question. Feasting on the social discontent that they do nothing to alleviate and rather often help cultivate, the far right seek to make hay from the fallout. A circular economy of reaction, or *metabolic grift*.

5
American Auto Apocalypse

Back in 2015, when Donald Trump exited the golden escalator in his eponymous tower and officially entered the populist zeitgeist, China was busy setting a course for what Xi Jinping called 'Ecological Civilization', built on 'quality productive forces'. Xi's announcement of a Made in China 2025 plan – which centred on shifting the economy towards high-value-added sectors, including electric vehicles (EVs) – is said to have 'alarmed' Washington.[1] The national 'alarm' could be read as an early tremor in a deepening crisis of American fossil hegemony, set off by the global shift towards decarbonisation and the rise of China as the dominant producer of clean technology. Here we turn to a core element of the so-called energy transition, the electrification of automobility, which sets into motion our broader exploration of the climate politics of Trump's third presidential campaign and the beginning of his second term.

The electric vehicle was a critical component of the crisis narrative circulated by Trump and those loyal to him in the lead-up to the 2024 election. While nativism remained the core of the Trumpist appeal and programme, the EV became part of a multifaceted crisis facing the nation. The partial phase-out of the combustion engine initiated under the Biden administration as well as California's own 2035 phase-out target, later rendered illegal by a GOP congressional majority, represented a genuine, if incipient, crisis for primitive fossil capital. The US is the world's leading consumer, producer and exporter of gasoline, with the transport sector in general and light-duty vehicles in particular constituting the largest source of national emissions. On top of this, the US's leading crude oil export industry is also deeply intertwined with the combustion engine, with a significant proportion of exported crude being refined into diesel and petrol for use in European automobiles, trucks and tractors.[2] Primitive fossil capital was clearly not going to take a transition to EVs lightly, and spent lavishly on Trump's and other Republican congressional campaigns.

Trump's campaign stitched these fossil-fuel interests into a carbon populist appeal, working to make a gestating mitigation crisis *feel* like a crisis for the American people too. Whereas the coal miner had been the star recruit from the ranks of the fossilised proletariat in Trump's first campaign, this time around it was the auto worker, now enrolled into an anti-EV polemic. Far from an incidental rotation in the cast of players, the US auto industry employs several orders of magnitude more people than the coal industry. Beyond employment, it remains critical to the 'defence industrial base', to most Americans' social reproduction and to the national imaginary. By 'lumping together' EVs with other phantasms of the far-right cinematic universe – immigrants, China, drugs, Black Lives Matter and the left more broadly – Trump stoked fears of cultural and demographic replacement, humiliation and national decline, with warnings that the signal American commodity is on the verge of a traitorous early retirement. As the enemy blurred into one, so too did the various moral panics on which MAGA politics had fed for the past decade. The moral panic cycle cashed out in one general panic, rendered in apocalyptic, blood-soaked rhetoric.

As if to concretise the narrative of the 'inverted crisis', this campaign strategy was translated into a mode of governance. Remarkable in speed and scale, a near-constant and escalating state of emergency became a hallmark of the second Trump administration. In its first hundred days, more emergencies were declared than in any year since the passage of the National Emergency Act in 1976. Coupled with the increasing subordination of Congress, Cabinet and the courts, the administration's formal declarations of emergency merged with a more general and informal category of extraordinary actions – from DOGE to the overturning of rules and procedures – to effectuate the Trumpist agenda of 'unitary' executive power. Framing the decarbonisation of mobility as part of a national crisis served as one technique for building consent for the use of exceptional measures and for a sweeping climate rollback. The Trumpist emergency state and the inverted crisis on which it depends represents the highest stage yet in the international process of fossil fascisation.

American bloodbath

Speaking at a Michigan auto parts manufacturer in September 2023, before a crowd ostensibly made up of striking auto workers, Donald Trump called for 'a revival of economic nationalism and our automobile-manufacturing lifeblood'. It was two weeks into the United Auto Workers', hereafter UAW, historic strike of the Big Three – GM, Ford and Stellantis – and Trump was in Michigan to inform those gathered that they were 'getting screwed', not by the companies, but by 'ultra left-wing globalism', 'radical left Marxists' and 'fascists', all of whom were in cahoots with Biden's EV push. The electric automobile transition, he told the crowd, would ensure that 'the future of the auto industry will be made in China', and would even 'destroy our country itself'. Such auto apocalypticism was a hallmark of the election and a central message in the critical swing state of Michigan, the heart of the US auto sector and the Trump–Vance campaign's second-most-frequented state. Polls showed that unionised auto workers had a negative view of Biden's clean-energy policies, and the UAW's own internal polling revealed that over a third of its membership supported Trump.[3] An anti-EV polemic – delivered across at least eight visits to the state in the closing months, including outside a proposed Chinese-backed battery plant – was a resonant wedge issue in Trump's successful efforts to flip the rust belt state red.

Despite addressing the crowd as if they themselves were striking UAW members and claiming to support their strike, the factory in which the September 2023 event was held was a *non-unionised* facility: Drake Enterprises, Inc., a medium-sized, family-owned company specialising in parts for heavy-duty trucks. The Trump campaign, it later emerged, paid the company $20,000 for hosting, thus securing a suitably industrial setting for the staging of its distorted show of solidarity. Since the facility did not employ unionised workers, some creative liberties needed to be taken in the construction of the *mise en scène*. UAW members in attendance at the invite-only event were recruited by local right-to-work activists and had their social media screened for sufficiently pro-Trump sentiment. On the day of the event, signs reading 'Union members for Trump' and 'Auto workers for Trump' were provided to individuals; some of whom, it turned out, were neither union members nor auto workers, but workers of a supplying firm. The event, in short, was a live-action deepfake, a manifestation of what Alberto Toscano describes as late fascism's 'ersatz' class politics.[4] A similar incident occurred during an October 2024 visit to the state by Vance, at which supporters wore 'Auto workers for Trump' T-shirts, despite not being auto workers. Of Trump's workerist veneer, one Michigan labour activist wrote discerningly that 'scabbing has truly never looked so desperate'.[5]

In their classic post-war study of American agitation, Leo Löwenthal and Norbert Guterman compare fascist demagoguery to two types of bad medical advice, dermatological and psychoanalytical. Both convey the same argument: 'For all his emphasis on and expression of discontent, the agitator functions objectively to perpetuate the conditions that give rise to that discontent.'[6] In the first metaphor, a patient with a skin disease is advised to give in to their destructive impulse to pick, claw at and otherwise disturb the affected area. Like an itchy scab, 'the agitator says keep scratching'. This may provide temporary relief, but the effect of heeding this quack prescription, of course, is to worsen the problem. In the second therapeutic metaphor, agitation is seen as 'psychoanalysis in reverse', one of its chief functions being to undermine self-reflection and agency by manipulating and psychically binding followers to the leader.

While pretending to approve of the strike, Trump downplayed the importance of the workers' collective action, framing support for him as far more powerful, a guarantee of patriarchal protection from the EV transition. 'You're striking for wages,' he said to the crowd and (more importantly) to the television and social media audience, 'but you know your job's only going to be here for two years or three years.' A vote for him, on the other hand, would be far more consequential, 'the ultimate strike against the globalist class'. Trump's ersatz class politics functioned like trade unionism in reverse: what mattered was not workers' collective power, but their symbolic role in a narrative positioning Trump as the only figure capable of confronting a mythic globalist class and the existential threat posed by the EV transition.

Two other analytical concepts used by Löwenthal and Guterman are relevant here: 'rehearsal' and 'lumping together'. The 'basic function of modern agitation', they write, is a 'rehearsal' of violence, whose 'most effective though indirect method' is the routine depiction of the enemy as a mortal threat to the 'people', an effort in which 'images of blood and violence' are central.[7] By vividly describing the enemy's ruthlessness in lurid detail, violence comes to be seen not only as necessary, but as fundamentally 'retaliatory', and therefore just. The enemy is multifaceted, interconnected through the conspiracist 'lumping-together device', a rhetorical technique that 'blurs the distinctions between all the enemy groups'.[8] It is this phantasmatic amalgam against which the agitator rages, though in practice the prime target tends to be the most vulnerable: in Löwenthal and Guterman's time, the Jew and, today, the migrant. 'The verbal fury of the agitator', they write, 'is only a rehearsal of real fury.'[9]

These concepts help make sense of the blood-soaked language Trump used to describe not only the electric mobility transition but the broader coalition of enemies to which it was linked. In his simulated union town hall, he warned against Biden's support for 'every single blood-sucking, globalist attack on US auto workers', saying the EV mandate is an 'assassination' of the industry and a 'hit job on Michigan and on Detroit'. Internally, a corrupt elite 'suck the life, wealth, and blood out of this country', while externally, 'other countries rape and pillage our jobs and our wealth'. Although Biden's EV policies were not the sole reason given for the impending devastation, they routinely coupled with a range of other causes and culprits on the campaign trail – 'radical left Democrats', 'Marxists and crazy people', 'fascists' – blending incomprehensibly into one another. Such bogeymen 'live like vermin within the confines of our country' and 'lie and steal and cheat on elections', and must therefore be 'rooted out', as Trump told a crowd in his 11 November 2023 Veterans' Day speech.

Stuart Hall and colleagues, less focused on specific agitators and more on broad-based 'signification spirals' within the mass media, refer to this as 'convergence', the second escalating mechanism in addition to violence thresholds through which 'crusading' occurs – that is, a process in which discrete moral panics proliferate with more intensity and frequency, while being 'mapped together' into a 'general panic about social order'.[10] At this point in the discursive convergence of many moral panics, 'the enemy is lurking everywhere'. Crusading thus culminates in a law and order campaign in which 'the crisis appears in its most abstract form: as a "general conspiracy"'. It is '"the crisis" – but in the disguise of Armageddon'.[11]

As a consummate performer of such convergences, Trump moves effortlessly between a variety of enemies and myriad scenes of betrayal, from 'open borders' and 'migrant crime' at one moment to soliloquies lambasting electric vehicles the next, mapping them together in one general conspiracy – in this case, the inverted American crisis.

On 16 March 2024, Trump was in Ohio to stump for Senator (then candidate) Bernie Moreno, former owner of a car dealership chain and one of the wealthiest members of Congress. The first half hour of the speech was pure border panic, including the recitation of song lyrics in which an injured snake is nursed back to health through the 'milk and honey' of a caring but foolish woman, only to then kill her with his poisonous bite. Linking the fable to migration, Trump concluded that 'we're taking in snakes', thus providing a novel variation on the theme of 'blood poisoning' that was a leitmotif of the campaign – immigrants are 'poisoning the blood of our country', he had stated some months before, while repeatedly forcing an association between migrants, drugs and overdose deaths. In the Ohio speech, this paradoxically biblical-yet-nihilistic 'devaluation of values' led straight to a call for 'the largest domestic deportation operation in American history'.[12] Moments later, Trump pivoted to EVs, championing a 100 per cent tariff on Chinese automobiles manufactured in Mexico, without which the nation would be in peril. 'Now, if I don't get elected, it's going to be a bloodbath for the whole country,' he prophesied.

Although 'bloodbath' is not an uncommon journalistic term used to denote mass layoffs or fierce competition rapidly eroding market share, the Trump campaign wedded industrial bloodbath to a broader crisis narrative. Days after the remarks in Ohio, the Republican National Committee created a website, BidenBloodbath.com, featuring a timeline of the 'border crisis' alongside details of fentanyl deaths and 'migrant crime' in various swing states. Weeks later, at a press conference in Grand Rapids, Michigan, Trump, flanked by police officers, declared that he would put an end to 'Biden's border bloodbath', eliciting applause from those beside him. While Michigan, thousands of miles from the southern border, might appear an odd state in which to deliver such a message, Trump deployed a spatial lumping-together device to argue that 'every state is now a border state. Every town is now a border town, because Joe Biden has brought the carnage and chaos and killing from all over the world, and dumped it straight into our backyards'.

The convergence of automotive and migrant bloodbaths was concretised in the 'One Big Beautiful Bill Act' (BBB), which decimated EV, battery and renewable policy support, while providing unprecedented funding increases to Immigration and Customs Enforcement (ICE), rendering it larger than most nations' military budgets. This convergence materialised further still in the physical practice of immigration raids, within which the automobile became a ubiquitous set piece. Traffic stops, often racialised and violent, were already intertwined with the carceral system, including immigration enforcement, serving as a 'primary gateway to mass incarceration'.[13] Empowered by the BBB, ICE has intensified these patterns of racialised road terror: videos circulated of masked agents taking sledgehammers to car windows; high-speed chases and collisions with agency cars; unmarked SUVs ramming vehicles belonging to immigration activists; in one case, an individual tried to flee, to which immigration officers responded by opening fire; in another, an ICE agent shot and killed a fleeing migrant; multiple others were killed by highway traffic as they fled on foot. As one commentator observed, the 'Trump administration's crackdown on illegal immigration often looks like a car crash. Literally.'[14]

The largest single ICE raid in terms of personnel and arrests, however, occurred on 4 September 2025 at Hyundai's EV battery plant in Georgia, the biggest capital investment in the state's history and one of the highest-profile firms ever subjected to such an operation. Border czar Tom Homan explained that it was about protecting American workers from 'illegal aliens' driving down wages. Nearly 500 workers, mostly Korean nationals brought over by Hyundai to fill skill shortages in the US's underdeveloped clean-tech labour market, were handcuffed, ordered by ICE agents onto buses, and held in detention for days. Given the performative excess of Trump's campaign and governance style, the setting was unlikely a fluke. More probably, the image of racialised and specifically East Asian EV workers in handcuffs was calculated to communicate a campaign promise kept: the furious verbal rehearsal – the migrant-auto bloodbath – reached its violent realisation. Yet any satisfaction drawn from such a spectacle must necessarily be temporary, a 'swindle of fulfillment' leading only to further escalating cycles of agitation and collective sadism.[15] In this death drive of decivilising passions, the MAGA base rides shotgun, 'watchdogs of order' enjoying vicarious power over the migrant.[16]

Eternal combustion engine

Accompanying Trump's apocalyptic defence of the combustion engine was a phalanx of civil society groups with connections to the fossil and auto industries, a powerful right wing media ecosystem, and state-level politicians and governments – a broad alliance that cohered first in opposition to Obama-era fuel economy and climate policy. This included corporate grass-roots campaigns by the American Fuel and Petrochemical Manufacturers (AFPM), a leading trade association of refiners. Motor gasoline is the sector's primary output, accounting for about half of the total, and so any diminishment in gasoline consumption is a direct hit to the bottom-line of AFPM's member firms. Their Energy4Us campaign strongly endorsed Trump in his first term, blanketing social media with ad campaigns calling on Americans to 'Support our president's car freedom agenda!' Auto firms, too, lobbied to reverse Obama's fuel economy standards, even as they cheated them in practice.[17] Although the official rule change did not arrive until 2020, this repeal was estimated to have been the single biggest climate rollback of Trump's first administration, permitting almost a billion tons more CO_2 over the lifetime of the relevant vehicles, equivalent to the annual emissions of a medium-sized country.[18]

When Biden took office the next year and announced his intention to restore and exceed the Obama standard, Trump was reportedly 'infuriated' – anger which would come to fuel a more aggressive rollback in his second term.[19] Delivered in March 2024 and described by the *New York Times* as 'one of the most significant climate regulations in the nation's history', the Biden administration's emissions standards for model years 2027–32, were, in fact, rather generous towards Detroit.[20] A year prior, the Environmental Protection Agency (EPA) had indicated that the forthcoming regulations would require EVs to make up two thirds of all new sales by 2032. Both auto capital and the UAW successfully pressured Biden to weaken these targets down to as little as one third, while also giving a larger role to hybrids. Despite getting their way, auto capital, as represented by the Alliance for Automotive Innovation, called it 'the ragged edge of achievable', meaning that these goals were possible only if 'everything … goes just right'.[21] Speaking more bluntly, head of the alliance John Bozzella admitted that 'most of the revenue and returns today are made from internal combustion engines'.[22] As one stark example of the truth of this, Ford's bottom line for 2024 saw its electric division lose about as much as its combustion engine division gained. In other words, profitability, not achievability, was the main driver of the auto alliance's advocacy. For his part, despite the concessions to auto capital, Trump began falsely referring to the Biden administration's emissions standards as an EV 'mandate', opposing them outright. Meanwhile, in fundraisers with oil and gas executives, Trump vowed to eliminate Biden's EV and other climate policies in exchange for a quid pro quo of a billion dollars in campaign contributions, ultimately settling for a paltry quarter billion spent across the election cycle. The AFPM, for its part, launched a new campaign urging citizens to mobilize to 'keep the government's hands off our cars', targeting swing state voters with a multimillion-dollar ad ambush.

Biden's approach to transport electrification was clearly not primarily ecological in orientation, but mainly premised on industrial renewal and out-competing China in the clean-tech sector, while promising to lock in automobile dependency, even if partially electrified, for another generation. His first major legislation, 2021's bipartisan Infrastructure Investment and Jobs Act, was a $550 billion infrastructure package with three quarters allocated to roads and bridges, after which only a paltry 4 per cent was left over for public transit.[23] Then, after a remarkable three-decade period in which the American denialist apparatus successfully forestalled federal climate policy, when it did finally come in the form of 2022's Inflation Reduction Act (IRA), it was significantly weakened. The IRA's loans, subsidies and tax incentives for renewables, EVs and battery production were counterbalanced by approval of the Mountain Valley natural gas pipeline and a quid pro quo by which any new offshore wind would require the simultaneous, or in some cases pre-emptive, issuing of oil and gas leases.[24] Thus, as

Brett Christophers emphasised in his post-mortem for Bidenomics, the IRA might have had a 'more beneficial effect for fossil fuel production than renewables development'.[25] The standard framing of the IRA as climate legislation, then, should be treated with caution.

From the beginning of the process of developing the IRA, leading MAGA mouthpiece Tucker Carlson spun out racist interpretations. 'Just as their supporters did in Minneapolis last summer', Carlson said, referring to the Black Lives Matter protest wave of 2020, 'the Biden people are looting the country' via their infrastructure plan. With respect to the IRA's partial funding for highway removal in select urban areas – particularly where highways have been intentionally routed through neighbourhoods of colour, often as de facto forms of racial segregation – Carlson mocked the very idea that a highway could have been designed in a racist way. 'Presumably,' he teased, 'the highway is sexist, transphobic and dog whistles for white supremacy, not to mention QAnon insurrections. Gotta tear it down.' Carlson's ham-fisted satire of a woke critique served to stitch road defence into a broader reactionary tapestry – shoring up, constructing and inoculating a fossilised common sense against a deranged and fanatical left seeking to ransack the roads. When the actual text of the IRA was finalised, Carlson heightened his rhetoric further, characterising environmental justice initiatives, such as the $30 billion in EPA funding earmarked for disadvantaged communities, as spreading 'race hate', comparing it to Nazism. The Heritage Foundation, then busy cooking up the soon-to-be-infamous Project 2025, came out against the IRA in similar terms, dubbing it 'green fascism'. In reality, many hoped the IRA would serve as something of a 'green' anti-fascist tool by winning disaffected parts of the electorate over to the Democratic camp, or at the very least insulating clean-tech investment from future Republican rollbacks.

Ultimately, the Republican Party – whether at the state or federal level – is united around prolonging the life of the internal combustion engine so long as it remains profitable. Notwithstanding the difficulties associated with tariffs and the industry's long-term competitive weakness in the EV sector, Trump 2.0 has extended the lifespan of the combustion engine well beyond previously anticipated expiry dates. Ford CEO Jim Farley gleefully calls the petro-revival 'a multibillion-dollar opportunity'.[26] Indeed, since Trump's return, the Big Three have announced billions in new investment for petrol and hybrid models; laid off thousands at EV plants and walked back promises of full electric conversion at several others; wilfully incurred billions in costs associated with backing out of EV-related plans and supplier contracts (in other words, spent billions of dollars to *not* decarbonise); and, finally, gassed up with MAGA momentum, intensified their pressure on Brussels and Ottawa to backtrack on 2035 combustion engine phase-outs. By late 2025 and early 2026, these efforts had borne fruit, with both capitals scrapping their targets. More minor yet indicative changes include a phase-out of engine start-stop technology and the restoration of inefficient product lines like the obnoxiously loud Hemi V8 engine for Ram pickups, now adorned with a 'protest badge' for those loyal customers who railed against its discontinuation.

The strategies and consequences of Trumpist climate obstructionism can be gleaned from GM's shifting gears during Trump's second term. Where obstruction can generally be understood as an effort to project the economic power of fossil capital into the spheres of culture and politics, GM's U-turn on the combustion engine follows a more circuitous path, responsive to increasingly muscular industrial policy.[27] Bidenomics was an effort to render EVs profitable, effectively purchasing auto capital's consent for the transition through the immense financial power of the US state. Breaking ranks with the other legacy auto makers, GM declared its intention to go beyond federal targets, opting instead for California's 2035 phase-out, which many Democratic states were set to follow, thus positioning itself ahead of what they saw as the inevitable EV curve. Trump's return would force a recalibration. Without federal tax breaks, loans and subsidies for EVs, GM reverted to obstructionism, investing billions into engine plants and mobilising employees to pressure lawmakers into repealing the California waiver – a special right of the state (pre-Trump 2.0) to impose more stringent emissions regulations than the federal standard.[28] As the firm's financial officer spoke of the internal combustion engine, the 'tail is now fatter and longer than anybody ever thought it was going to be. That is going to be a much bigger cash flow engine.'[29]

As in his first go around, Trump 2.0 can be understood as a merger of the denialist apparatus and the state.[30] The GM case suggests that this convergence, in turn, alters the strategies of leading firms beyond the oil and gas sector. The denialist state, through deregulation and anti-green austerity, creates a permission and incentive structure that drags along other sectors of the economy towards fossil fuel interests, now capable of asserting economy-wide leadership. Here, the economic power of the fossil fuel industry expresses itself in politics, which in turn expresses itself back through the economy, rejigging the incentives of other fractions, who, in turn, pressure the state according to their newly fossil-aligned accumulation strategies. Obstruction begets obstruction.

Enter Musk, stage right

Elon Musk's role in the 2024 election marked a new phase in the alignment between tech billionaires and far-right politics. Musk became a central force behind Trump's third presidential run, pouring in hundreds of millions of dollars, using X's algorithms to amplify MAGA content and personally dedicating himself to the campaign. Echoing hard-right talking points, Musk warned that a Biden victory would mark 'the last free election', claiming that Democrats would 'legalise' millions of migrants to end American democracy and bring about 'the end of Western civilization'. Despite Musk's sharp pivot to far-right culture wars in the years preceding the election, the question remains, why did the CEO and largest shareholder of the country's leading EV producer – a company that has survived largely as a result of Democratic policy, especially California's zero-emission credits – throw his full weight behind a campaign that had demonised the very products on which his fortune had been built?

The contradictions are dizzying. While the legacy auto makers lobbied the Biden administration for weaker tailpipe regulations, as recently as July 2023 Tesla lobbied the EPA to phase out most combustion engine vehicles by 2035, arguing that doing so was 'essential' to solve 'the escalating climate crisis'.[31] And in February 2024, when the EPA was set to pass its watered-down rules, Tesla lobbied the agency to grant California its waiver to exceed the forthcoming national standards. In other words, over roughly the same period as its CEO was making a hard-right turn to a brand of politics dead set against climate action in every imaginable way, Tesla was quietly pressuring the Biden administration to aggressively decarbonise automobility.

It was long believed that 'green' capital could (eventually) constitute a sufficiently powerful lobby to steer the state away from the most egregious demands of the oil and gas industry.[32] The thing about green capital, however, is that it does not really exist, at least not in pure form or at sufficient scale to provide such a counterweight.[33] If Tesla's executives and largest shareholders were all purely invested in clean tech, then we could perhaps consider Tesla a pure instantiation of 'green capital'. The reality, however, is far messier: an EV company whose revenues and profits are dependent on the immense degree of pollution spewed by other carmakers, led by a libertarian whose pragmatic side never misses a chance for government handouts; a man whose myriad other business and personal interests can and do come into conflict with the more narrow interests of Tesla; a company whose 'interests', however, appear to be shifting as it becomes more integrated with other elements of the CEO's multifaceted empire. Narrowly understood, Tesla's interests may be reflected well in their lobbying records: in the short term, the firm wants stricter emissions standards, ideally stringent enough to set up legacy auto makers to fail, such that they are forced to purchase more credits from Tesla. In the medium term, the firm wants a full combustion engine phase-out, confident it can maintain a substantial portion of its sector-leading market share. Tesla, however, cannot be understood in isolation from Musk's broader business empire.

A Trump presidency offered clear material incentives to Musk, who stood to gain through the intertwined interests of his flagship firms, Tesla and SpaceX. (After mergers in early 2025 and 2026, his other two major firms, xAI and X, are now subsidiaries of SpaceX, making it the world's most valuable private company.) Trump's militarism presented a possible boon to SpaceX, to which roughly half of Musk's net worth was then tied. A Trump administration, as Musk well knew, could free up subsidies for the SpaceX subsidiary Starlink, a satellite internet company that had been excluded from a Biden-era programme to provide rural internet. SpaceX's rocket programme also benefited from Trump's militarisation of space. And in their first public conversation livestreamed on X, Musk had already noted that Trump could undercut the labour movement, which was increasingly seen as a threat to Tesla, particularly starting in autumn 2023, from the wave of strikes across Scandinavia in solidarity with Swedish union IF Metall to the UAW's victory in that year's stand-up strike and subsequent declaration of intent to organise Tesla workers. Indeed, Tesla's anti-union politics were the reason the Biden administration snubbed the firm at a 2021 EV summit with the UAW and the Big Three, further driving a wedge between Musk and the Democrats.

MAGA politics also arguably served as a more logical political cross-branding exercise to accompany the latest addition to Tesla's line-up, the Cybertruck. At a late 2023 event commemorating the first deliveries of this stainless steel, *Blade Runner*–inspired monstrosity, Musk screened a video of the truck withstanding machine gun fire and declared, the 'apocalypse can come along any moment, and here at Tesla we have the best in apocalypse technology'. The Cybertruck is perfectly pitched towards the mainstreaming of prepper culture and the associated boom in the 'panic industry'.[34] Already increasingly popular among the libertarian right through the Obama years, this trend has intensified since 6 January, not to mention the pandemic and multiple global conflagrations. Every prepper must have a 'bug-out location', where one can go when SHTF (shit hits the fan), and in order to get to one's safe space, one needs a reliable 'bug-out vehicle'. The Cybertruck appears to be something of a *faux* bug-out vehicle for influencers, the very online and finance and tech bros; real preppers would choose something more practical, while the bourgeoisie proper keep private planes and helicopters fuelled up to avoid traffic. If the SUV was already an expression of the 'automotive subjectivity of neoliberal capitalism', in which personal safety is elevated above that of society in a zero-sum competition, then the Cybertruck arguably expresses the radicalisation of the neoliberal subject in a social and ecological context in which breakdown and collapse have become a widely shared structure of feeling.[35] Marketed as 'built for any planet' and equipped with bulletproof siding and a 'bioweapon defence mode', the Cybertruck promises that its owners will be 'beyond prepared', offering a sort of climate adaptation for climate denialists.

Equally important in explaining Musk's unparalleled support of Trump 2.0 is the radicalisation of Silicon Valley and its related pursuit of artificial intelligence, a boom which Musk intends to lead with his own entry into the field via xAI, with which Tesla is increasingly integrated. Self-driving cars and humanoid robots are now held out by Musk as the main promise for long-term profitability after the advance of Chinese EV companies like BYD set back its ambitions to produce a mass-market, low-cost vehicle, long held out as the rational kernel of Tesla's overvaluation. Both of these technologies are underpinned by AI, and so personnel and financial interlocks have grown between his breadwinner and the new start-up. In midsummer of 2024, for example, Musk sought approval from the Tesla board to invest $5 billion in xAI, while diverting $500 million worth of scarce NVidia chips away from the carmaker. The relationship is reciprocal, too, with Musk telling shareholders, 'Tesla is learning quite a bit from xAI.'[36]

And still, it is hardly clear that the opportunity a Trump administration represented to head off regulation in both autonomy and AI has drastically outweighed the reputational and (de)regulatory costs, nor the benefits that could have been expected from another Democratic administration. Musk's MAGAfication has led to a predictable consumer backlash that has dramatically lowered Tesla sales in Europe and elsewhere. Perhaps more significantly, Trump's Big Beautiful Bill eliminated one of Tesla's most profitable lines of business: the sale of regulatory credits. Since fines for violating fuel economy standards have been eliminated, there is no incentive to purchase offsets, thus evaporating what has historically been one of Tesla's leading profit streams.[37] While framing his split with Trump as a principled stand against the BBB's contribution to public debt, Musk's ultimate split from Trump in summer 2025 may have centred more on this under-reported blow to his business. Although the patrimonial control enjoyed by Musk across his empire resonated with Trump's authoritarian 'unitary' executive, the contradictions of the alliance were too great. In the end, there could only be one *pater*.

A variety of explanations for Musk's political pivot have been floated across the political feuilleton and podcast sphere: his attraction to risk, transphobia against his daughter, hurt feelings at the EV summit snub, childhood bullying and paternal abuse, use of ketamine and other drugs and his penchant for conspiracy theories becoming radicalised through obsessive use of Twitter during the pandemic. Ideological radicalisation across his social milieu, however, seems critical. Musk may be the most visible and vocal of the lot, but he is part of a longer-standing rightward radicalisation in Silicon Valley, animated by neoreactionary 'dark enlightenment' thinking, eugenics, long-termism, techno-optimism and accelerationism. Musk's PayPal co-founder and chairman of Palantir, Peter Thiel, has been at the vanguard of this fraction since at least the first Trump administration, and his efforts in cultivating new political talent have paid off spectacularly in the rise of JD Vance, a protege of Thiel's whose selection as Trump's running mate Musk had personally lobbied for. Thiel, for his part, explicitly rejects democracy and has close ties with the principal neoreactionary thinker Curtis Yarvin, aka Mencius Moldbug, who wishes for the US to be governed as a private corporation, with CEO kings ruling over citadels of consumption. In the work of this techno-fascist organic intellectual, the patrimonial structure of the tech firm is self-consciously extended as an authoritarian theory of state, closely paralleling that promoted by Project 2025.

In October 2023, Marc Andreessen – another Silicon Valley billionaire, friend of and investor in Musk and his assorted ventures, and prominent Trump donor – published 'The Techno-optimist Manifesto', which serves as an insight into the mood of the Silicon Valley right.[38] The gist of the text is that tech CEOs should be able to do whatever they want without any government regulation, thereby unleashing AI, a technology that will save untold millions of lives and which, therefore, would be murderous to delay. The manifesto situates itself in the same lineage as the Futurist manifesto, written by Filippo Tommaso Marinetti in 1909, a document widely seen as a cultural and intellectual precursor to Italian Fascism, in which fossil-powered machines and speed – expressed above all in the automobile – are championed as the harbingers of a new type of man, capable of triumphing over nature, women and 'competing races' to revivify the nation.[39] In his contemporary update, Andreessen quotes Marinetti favourably, naming him among the 'patron saints of techno-optimism'. Like Marinetti before him, Andreessen's manifesto has special praise for the 'eros of the car'.

Pedal to the metal

In May 2025, less than a year out from his four-month prison stint, leading MAGA strategist Steve Bannon reflected on the differences between Trump's first and second terms. The first term, he admitted, suffered from a general lack of preparation.[40] This time was different. In addition to using his *War Room* podcast to mobilise followers to take over positions within the local branches of the Republican Party, school boards and poll-monitoring positions (the so-called 'precinct strategy'), Bannon credited a vast ecosystem of right wing public intellectuals with designing the game plan for Trump 2.0, the 'overarching' efforts of the Heritage Foundation's Project 2025 foremost among them. If 'flooding the zone' was originally a tactic meant to distract political adversaries and the media with a blitz of sensationalist and reality-bending rhetoric, then in Trump's second presidency it would refer to a barrage of executive orders, policies and states of emergency. Despite his hostility towards Musk's business operations in China, Bannon praised DOGE as a complementary, blunt force blow to the deep state. That this plan could be put into authoritarian action – 'pedal to the metal on executive power', as one legal scholar put it – would suggest that Project 2025 merits a brief review.[41]

The Heritage Foundation has long been integral to the denialist apparatus, and so it is unsurprising that the climatic dimensions of this far-reaching effort to 'deconstruct the administrative state' are unambiguously pro-fossil fuel. The chapter on transport, for example, called for a rollback of fuel economy standards; the elimination of the California waiver; and an end to federal funding for sidewalks, bicycle lanes, hiking trails and public transit. Project 2025, however, is not only a suite of reactionary and anti-ecological policy proposals, though; its ambition is to reconstruct 'the American state from the ground up'.[42] Such efforts are based upon a theory of the 'unitary executive' that ordains the president with 'supreme, king-like powers' to dismiss agency heads; reclassify permanent civil employees as partisan appointees; and supersede the power of the purse, meaning that agencies and programmes approved by Congress may be defunded. The sprawling web of agencies constituting the federal government, then, become subject to the 'patrimonial' control of the president. Meanwhile, the conservative Supreme Court majority, stacked in Trump's first term, provides the exceptional state form with a judicial backstop.[43] As lower courts are expected to and often do rule against such executive overreach, the Supreme Court likewise assumes an exceptional form, increasingly using its emergency or 'shadow docket' to neutralise these decisions without the need for briefings, written decisions or the public scrutiny which these typically entail.

When the rubber of unitary executive theory hit the road in Trump's second term, the administration rapidly implemented the most significant of Project 2025's transport policies.[44] Trump 2.0 went after the California waiver in swift and innovative fashion. In his first term, Trump relied on using standard but lengthy procedure.

Accelerating the process this time at the behest of EPA head Lee Zeldin, congressional Republicans steamrolled the parliamentary rules that forbid their plans and rescinded the waiver with a simple majority.[45] Explaining the somewhat wonkish implications of this, Kate Aronoff writes that these acts of procedural malfeasance 'seem to have opened the floodgates for what can be accomplished using the Congressional Review Act', thus fundamentally transforming the Senate.[46] While perhaps arcane relative to the brazen disobedience of courts or the elimination of an entire agency (like USAID), such acts of parliamentary rule breaking are an effective weapon in the war on the administrative state.

Amidst the flurry of day-one executive orders, Trump declared two states of emergency, a seeming attempt to fulfil his pledge to be 'dictator for a day', close the border and 'drill, baby, drill'. He would go on to declare six more states of emergency in the first six months of his term, more than any recent president had in a comparable period, second only to himself in the latter part of his first term. The first was a state of emergency at the southern border, the second a 'national energy emergency'. The latter directed the heads of all government agencies to 'exercise any lawful emergency authorities available to them' to accelerate energy projects, specifically defining 'energy' in such a way as to exclude solar and wind. The US was said to be suffering from grid unreliability and insufficient power generation, representing a 'growing threat to the United States' prosperity and national security', despite the government's own research contradicting these claims. The declaration, noted legal commentators, created an 'unexplained and unwarranted sense of urgency around an invented crisis' – a pointed elaboration of how an inverted crisis functions.[47] Thanks to court decisions and legal limbo, however, the president has enjoyed relatively unrestrained power to define what a 'crisis' is. Hundreds of fossil fuel projects, from pipelines to natural gas terminals and power plants, are slated to be fast-tracked under the emergency declaration, overriding environmental protections and inverting climatic and biodiversity crises not just rhetorically, but in practice.

Spurred on by DOGE, the EPA cut thousands of staff, shuttered programmes and offices and transformed into an engine for enabling corporate emissions and pollution. On 12 March 2025, Zeldin announced thirty-one measures amounting to, in his own terms, 'the most consequential day of deregulation in US history'. Concretely, this included the rewriting of tailpipe emissions standards for light, medium and heavy-duty vehicles; the weakening of emissions reduction requirements for power plants; and the attempt to reverse the so-called 'endangerment finding', a legal document confirming the negative impacts of greenhouse gases. Despite controversy, in July 2025, with a heavy-duty truck dealership chosen as his venue, Zeldin announced he would initiate formal proceedings to revoke the finding. Some months later, the rule had been officially rescinded, thus eliminating the scientific basis for most US climate and clean-air regulations. As for the ostensible benefit the American public might see from all of this, Zeldin boasted, 'Today the green new scam ends', and touted how deregulation would make it 'more affordable to purchase a car, heat your home, and operate a business' – in short, an absolute inversion of the EPA's purpose.

Additional efforts of the fossilised unitary executive have run through the White House's Office of Management and Budget, led by Project 2025 architect Russell Vought. Referred to as the 'glue' between DOGE and the administration, Vought was successful in demanding the rescission of tens of billions in congressionally approved science funding, derided as 'woke'. In the context of the government shutdown in autumn 2025, the office cut an additional $8 billion in funding for hundreds of clean-energy projects, from battery plants to grid updates, targeted specifically in Democratic states. This was on top of the administration withholding $18 billion in federal funding for mega-projects like New York City's Second Avenue subway and the Hudson river tunnel for commuter rail.

Trump has not gone completely unimpeded in his use of emergency powers. His global trade and tariff policy, which relies on yet another invented state of emergency and included 25 per cent tariffs on all foreign-made automobiles, faced headwinds at the Supreme Court, a rare instance of pushback. Expressing the contradictory views of the court's conservative majority, Justice Neil Gorsuch worried that granting the president the right to dictate trade policy on the basis of imaginary emergencies could result in a different (i.e., Democratic) administration implementing 50 per cent tariffs on combustion engine vehicles under the pretence of a 'climate emergency'.

While executive orders, emergency powers and the subordination of Congress and courts define Trump 2.0, major policy changes, including climate rollbacks, have been implemented through regular channels: notably, the flagship One Big Beautiful Bill Act. Beyond the massive tax cuts to the rich, the elimination of health care for millions and the massive increases in immigration enforcement and military funding, the BBB stripped away solar and wind incentives, slashed fines for violating fuel economy standards and eased taxes for luxury vehicle loans, while taking a sledgehammer to electric vehicle tax credits and subsidies.

As a result of this deregulatory barrage, short-term EV adoption forecasts have been halved, EV-related investments have been reduced by as much as a third year-on-year, and billions more in planned investment have been cancelled. Investment in renewable energy shrank by 25 per cent in 2025 compared with the previous year, while $18 billion worth of projects were cancelled as of August 2025. Trump's war on wind and solar has posed difficulties for leading firms, with concerns around stranded assets and share price devaluation flipping from one side of the energy transition to the other – a spectre no longer haunting fossil companies, but renewables firms. Danish state-backed Ørsted, for example, the world's largest offshore wind producer, has been hit hard by the political sea change, evident in a record dip in its stock price that set off when Trump ordered a work stoppage on an almost completed offshore wind farm project near Rhode Island. The outlook for commercial solar, too, has become cloudy. The largest renewable energy company in the US, NextEra, has seen its ambitions to create North America's largest solar farm dashed by the Trump administration's cancellation. Zooming out, models forecast that by 2035 the BBB will have caused a drop of up to 40 per cent in EV adoption and a 70 per cent reduction in renewable energy additions to the grid.[48]

Trump 2.0 has also menaced global climate diplomacy, by both action and inaction. A day-one executive order saw the US once again withdraw from the Paris Agreement. As a result, American officials were decidedly absent among delegates of COP30 in Belém, Brazil, where Chinese EVs served as the official transport. The administration also withdrew from a variety of international climate finance agreements designed to aid developing countries with mitigation and adaptation. A defender of combustion engines of all shapes and sizes, the Trump administration likewise torpedoed a global agreement to reduce shipping emissions, which account for 3 per cent of global emissions, equivalent to the airline industry. To sink these ambitions, the administration reportedly threatened developing countries with tariffs, withdrawal of visa rights and the blacklisting of diplomats.

In *Policing the Crisis*, Hall and colleagues observed that a 'cycle of moral panics issues directly into a law-and-order society'.[49] Given the scale of the concocted crisis, a period of 'more than usual law' is required to 'set things right'. In other words, the intensification and convergence of many moral panics into one general crisis tends to cash out as a transformation in the role of the state. Yet it would be difficult to characterise Trump's return as a period of 'more than usual law' – in many respects, it manifests instead as a 'constitutional crisis', owing to numerous refusals of court orders as well as authoritarian disregard of basic rights and laws. Hall was careful to emphasise that he was not describing a transition into an exceptional form of state like fascism, but rather into an '"exceptional moment" in the "normal" form of the late capitalist state'.[50] In this exceptional moment, the state moves away from consent as the dominant mode of securing hegemony, and 'tilts' towards the 'pole of coercion'. The exceptionality of the law-and-order society thus consists not so much in the creation of new powers as in the increased use of 'coercive mechanisms and apparatuses already available within the normal repertoire of state power'.[51] Trump 2.0 marks a transition to a qualitatively different, exceptional form of state, where the discourse of crisis is now used to justify not only coercion, but large-scale authoritarian transformation – a movement of the inverted crisis as campaign rhetoric into a 'crisis form of state'.[52]

In the waning days of Trump's first administration, Joel Wainwright and Geoff Mann predicted that the cultivation of crisis narratives would play a critical role in any possible return of Trump. In their observation, climate denial was losing importance to the Trump project, and what increasingly mattered to its future success was instead the 'conjuring of an apocalyptic *political* climate'.[53] What we have shown is that climate denial retains its hold on the far right precisely through this apocalypticism. Since its arrival on the world stage, MAGA politics has been characterised by a ceaseless parade of folk devils. On the campaign, these were lumped together with climate politics more broadly, and the EV transition in particular, resulting in a convergence of all crises into one general crisis: the American auto apocalypse. Such carbon populism added a visceral quality to the threat posed by otherwise distant phantasms. In this great driving right show, the crisis motors straight into the heartland, where economic anxiety, Fordist nostalgia, fear of imperial decline at the hands of a rising (*communist*) China, 'gender ideology', 'antifa', 'cultural Marxism', 'critical race theory' and 'wokeism' are intertwined. Above all, the leader requested two simple actions: vote for me and enjoy your internal combustion engine. Without the leader's protection, what awaited the nation – from the universal border town to the industrial heartland – was nothing less than a 'bloodbath'.

The radicalisation in Trump's campaign language quickly materialised into radicalised statecraft. In shifting from rhetorical emergency to state of emergency, the former builds consent (however narrow) for the latter, providing a discursive and legal framework through which emergency powers can transform the state. To be sure, the existing constitutional order helps create the opening and even co-produces the constitutional crisis.[54] With a Supreme Court sympathetic to the unitary-executive theory, coupled with a pliant Congress and a rabid base, major hurdles to state transformation have been cleared. Taken to its logical outcome, the vision of Project 2025 points towards a permanent emergency state, converting presidential powers into something more dictatorial or monarchical, as envisioned by the organic intellectuals of the Silicon Valley fascists: the state is refashioned in the mould of a patrimonial tech start-up.

In *Policing the Crisis*, Stuart Hall and colleagues conceptualised the moral panic cycle as culminating in an exceptional moment in the normal functioning of the state. Trumpism has now gone well beyond this. It is not simply a turn away from consent and towards coercion. Though it relies on previously established emergency powers, the project is clear and deliberate in its aim to invent entirely new ones, whether through DOGE, the supersession of Congress or the disregard for existing laws and court decisions. It is not a period of 'more than usual law', as Hall described authoritarian populism. Rather, Trumpism hews closer to Paxton's description of fascism, which produces a 'sense of overwhelming crisis beyond the reach of any traditional solutions', and, as a real historic force, 'abandons democratic liberties and pursues with redemptive violence and without ethical or legal restraints goals of internal cleansing and external expansion'.[55] Trump 2.0 is, if anything, a period of *less* than usual law. As the most advanced stage yet of fossil fascisation, Trump 2.0 builds its power by forecasting both industrial and migration-fuelled bloodbaths, building justification for an emergency state required to protect the border and the combustion engine alike.

6
The Remigration Industrial Complex

'People will have to move to survive,' writes Gaia Vince in *Nomad Century*, referring not to a distant few but to nearly half of today's global population.[1] Over the next fifty years, rising temperatures and humidity levels are projected to make vast parts of the world dangerously uninhabitable for up to 1.5 billion people, forcing unfathomable numbers into the category of 'migrant' or 'refugee'.[2] In a world 2°C warmer than pre-industrial levels – a scenario the IPCC warns could arrive by the end of the century, if not earlier – the map of the habitable earth will be forcibly redrawn. The equatorial belt will become effectively unliveable for much of the year, and the mid-latitude zones will endure extreme seasonal stress. In this not improbable scenario, survival will depend on movement towards high-latitude regions and the poles. In a world without mitigation, humanity is pushed to the planetary margins.

As extreme weather events and mass displacement intensify in tandem, so too will the violence at the nexus of climate and border regimes. The far right has long supplemented the early denialist effort by primitive fossil capital to fabricate and disseminate a flood of lies about climate change. Increasingly it steps in to fill a cognitive vacuum created by the undeniable experience of an environmental crisis striking with ever-increasing speed and levels of destruction on the one hand and the lack of an adequate explanatory framework on the other. In this narrative architecture, climate-induced disasters are not evidence of systemic failure but symptoms of invasion, contamination and decline attributed to the Other. This dynamic lies at the heart of what we have called the inverted crisis: the problem is not climate change itself, but what certain actors seek to do about it. While much of this book has examined the far right's demonisation of such figures, here we argue that, as a logical extension of the inverted crisis and the global rise of ultranationalism, the primary scapegoat becomes the racialised Other, the immigrant.

In times of climate collapse and the consequent mass displacement, one of the far right's most enduring ideological fantasies turns fantastically plausible: the 'Great Replacement' is no longer just racist conspiracy theory about demographic change, but reinterpreted as a form of war. In *White Skin, Black Fuel*, it was argued that environmental crises increasingly serve as justification for violent and exclusionary regimes.[3] The far right frames the migrant as the harbinger, or even the root cause, of planetary disorder – a figure that must be contained, expelled or eliminated. With conspiratorial climate denial and racist paranoia crystallising in even the highest political offices, this trajectory suggests that as the world burns, borders will only harden. Taken together, the stage is set for an era of intensified border violence justified not in spite of climate breakdown, but because of it.

From economic stagnation to ecological destruction, this process is propelled by various dimensions of what we have called an emergent organic crisis. As material conditions deteriorate across multiple planes, and as the far right intensifies an already diffuse climate of social anxiety, both migration and 'green' politics are framed as twin drivers of national decline. In Germany, this discourse has been refined and institutionalised, as the far right has long centred its rhetoric on the fear of national economic collapse. With warnings about 'deindustrialisation' and panic about an automobile industry challenged by the EV transition, the far right has become increasingly successful at converting popular fears into electoral gains. A coherent industrial policy, of course, rarely comes along with its political agitation. Instead, popular anxieties over economic erosion and geopolitical decline are redirected and projected onto a host of actors and channelled towards a range of demands, above all: 'remigration'.

The long shadow of blood and soil

In an early bid to push the link between climate breakdown and mass migration into the mainstream, the youth wing of the AfD sent an open letter to the parent party in 2019, urging it to stop officially denying climate change.[4] The Junge Alternative had not suddenly embraced climate science; rather, it observed that 'climate change and environmental protection' had been a decisive topic for the electorate in the European elections, in which far-right parties gained fewer votes than they had anticipated. Seeking to appropriate the momentum of left-wing climate politics for reactionary ends, the youth group adopted a right–Gramscian strategy: seize the climate narrative, recast its meaning and redirect its ideological thrust.[5] Hence the Junge Alternative's proposal for how the parent party should *address* climate change rather than denying it: tie development aid to the adaptation of a one-child policy. In an inverted twist to anthropogenic climate change, the open letter details that the human impact on the climate is 'undisputed', but that it is overpopulation in the global South – not fossil fuel emissions, lavish lifestyles or capitalist interests – that lies at the heart of this 'impact'. In its programme *Youth Taking the Lead!*, adopted in 2022, the Junge Alternative extended its Great Replacement take on climate change: 'Climatic changes must not lead to a further weakening of European border protection. We reject any further immigration under the label of so-called climate flight.'[6]

The ecological imaginary of the Junge Alternative is not novel. Rather it draws from a long lineage in which the racialised Other is cast as a threat to ecosystems or a mystified notion of 'nature'. At its most virulent, the blood and soil ideology, a central tenet of Nazi thought, fused ethnic nationalism ('blood') with agrarian romanticism ('soil').[7] By promoting the idea that a people's racial purity was deeply tied to their ancestral land, only those of 'pure' German blood were deemed legitimate stewards of the land, while others – particularly Jews, Roma and Slavic peoples – were cast as alien intruders defiling the natural order. Following the war, the weddedness of race, nature and territory was cast to the back chambers as it remained part and parcel of many now-firewalled far-right ideologies. As the overt expression of eco-fascist thought carried high social costs for decades, it was largely relegated to the margins: fringe movements and demagogues operating outside the bounds of political respectability.[8] The terrain on which the far right operates today, however, is shifting. With it, more radical elements of its ideology are revived and absorbed into collective sense making. What once remained confined to the manifestos of mass shooters and neo-Nazi forums is creeping closer into the mainstream, where signs of a final inversion – casting migrants as the culprits of climate disasters – are already visible.

As the largest and longest wildfire ever documented in the European Union blazed in the Evros region of Greece for over two weeks in 2023, the far right opened a new chapter in the inversion of natural disasters. The flames were indeed framed as 'anthropogenic' – not, however, caused by burning fossil fuels, but by 'illegal migrants'. What began as a fringe conspiracy quickly leapt into mainstream discourse, spreading through social media and sensationalist outlets even faster than the fire cut its path through the region. But it was the narrative's adoption by senior officials that signalled its deeper political traction. Twelve days into the blaze, Greek Prime Minister Kyriakos Mitsotakis stood before the parliament and, without offering evidence, claimed, 'It is almost certain the causes were man-made. And it is also almost certain that this fire started on routes that are often used by illegal migrants.'[9] Spurred by the violent rhetoric of far-right MPs, who declared that 'we are at war' as 'illegal immigrants have entered in a coordinated manner and ignited more than 10 fires', ad hoc militias emerged, with teenagers taking to the streets on scooters to patrol. Others took far more violent measures. Three local men – described by the Alexandroupolis deputy prosecutor as 'bounty hunters driven by racist motives' – abducted thirteen migrants, locked them in a trailer, and drove them towards the border.[10] In fact, all twenty victims who perished in the fires were identified as migrants en route. In this double bind, migrants became both the first to die in the flames and the first to be blamed for them – victims not only of climate collapse, but of a denialism that displaces responsibility through violence and turns the most vulnerable into culprits.

Evros's particular geopolitical location, however, suggests that this kind of victim–perpetrator reversal will continue to unfold in the region. Straddling the border between Greece and Turkey, the Evros region has become a focal point of both migration flows and border militarisation. As a key corridor for unauthorised entry into the European Union, the Evros river is a hotspot of crossings by refugees and migrants. Though often labelled a 'natural' border, the river has been aggressively securitised – transformed into a militarised buffer zone lined with fences, watchtowers and patrol roads, while observers, researchers and medical professionals are prevented entry.[11] The violent defence of the EU borders against a foreign Other radiates into the surrounding region, where anti-immigration sentiments and attacks on refugees have long histories. What makes Evros prefigurative of future politics, then, is how the prominence of migration flows enables political actors to intensify narratives that scapegoat migrants for climate-related disasters.

Greece is no isolated case. Rather, each climate-induced event seems to spawn its own narrative. In the wake of Hurricane Helene, which struck western North Carolina in September 2024, conspiracy theories quickly spread online, some of which falsely claimed that federal disaster relief funds were being redirected to support undocumented migrants. Both Donald Trump and Elon Musk amplified these baseless allegations on X, propelling them into the mainstream. Following the deadly floods in Valencia, Spain, that killed over 200 people in late 2024, there was 'a storm of denialist mud and reactionary sludge' online, amplified by the far-right party Vox, accusing the government, progressive NGOs and environmentalists of maleficence; the latter supposedly dismantled Franco-era dams that would have contained the deluge.[12] The winds of unreason also fanned the California wildfires in the early months of 2025: Democrats deliberately withholding water, globalists using energy weapons, the LA fire department being run by a 'DEI hire'.[13] Alex Jones offered a more idiosyncratic interpretation: 'It's official, LA is going to be turned into an AI-run fifteen-minute city, open-air prison!' But first 'they' had to burn down what was already there. In a political landscape gripped by denialism, conspiracy theories rush in to explain the otherwise inexplicable: why disasters are intensifying, why lives and livelihoods are increasingly lost. The material crisis and its causes are denied and downplayed, with panic and social anxiety redirected towards scapegoats – all these fires and floods must be the fault of those who wish harm upon the nation.

While the overt accusation of migrants as the culprits of climate catastrophes has not yet become routine, the far right is cultivating a more diffuse climate of national anxiety, in which migration and 'green' politics are framed as twin drivers of economic decline. Beneath the surface of moral panics and lurid conspiracies lies a deeper struggle over who gets to move, who is forced to stay and who is excluded altogether. Already in its 2021 election campaign, the AfD made this logic explicit. One poster, plastered across cities, read, 'Will I soon no longer be allowed to go to Crete, Greta?' above an image of a sad blonde girl staring out of a rain-speckled window. It tapped into anxieties that personal travel freedoms might be permanently curtailed in the so-called new normal. Another poster, meanwhile, proclaimed that 'Cologne, Kassel and Konstanz can't take any more Kabul'. This dual messaging reveals the racism underlying the imperial mode of living: mobility as birthright of the global North, but always disputed for those beyond.

It is in this context that the far right's defence of fossil fuels reveals itself as more than nostalgic protectionism for a fading world order. It is an active intervention in what Mimi Sheller calls the global 'politics and power relations of (im)mobilities' – the entanglement of migration, climate crisis and urban life.[14] The fossil-fuelled mobility regime is built on a hierarchy that grants freedom of movement to some while violently denying it to others. As the climate crisis accelerates, this paradox sharpens. The 'freedom to move' for some threatens the very 'freedom to stay' for many: to live securely in a place one calls home.[15] In this context, the far right's mobility politics mutate into something even more extreme: a project that seeks not only to restrict movement, but to reverse it.

The globalisation of remigration

At a clandestine meeting in Potsdam in November 2023, twenty-two far-right activists, AfD elected representatives and business owners forged a strategy for the deportation of roughly 20 million people living in Germany.[16] Throughout this gathering, the concept of 'remigration' was advanced, pitched to be an essential element of a wider project of regime change from the right. The use of the term 'remigration' is itself a cynical inversion. Once used to describe the return of Germans who had fled Nazi persecution after the Second World War – especially Jews, Roma, socialists and queer people – the word has now been repurposed to normalise one of the most extensive racist agendas since the fall of the Third Reich. Branded by its most fervent advocate and the leader of Austria's Identitarian Movement, Martin Sellner, the 'remigration' scheme marks the far right's effort to turn a conspiratorial belief in the Great Replacement and racist desire for ethnic purity into actionable policy: reversing demographic shifts through systemic expulsion. Remigration, in the words of the French New Right, aims at 'inverting the migratory flows' through mass deportation.

When, in January 2024, the news of the Potsdam meeting broke, massive protests erupted across Germany. With over 3 million participants at more than 1,200 rallies, it became the largest wave of demonstrations in the history of the Federal Republic. For a brief moment, it seemed as if the AfD – then the largest opposition party in the parliament – might be rattled by the unmasking of its extremist core. Rather than disavow the plans, however, high-ranking officials of the party leaned in and embraced them, slowly moving sceptics within its own ranks further to the right. 'Yes,' members of the parliament declared, 'remigration has always been central to our party's DNA.'[17] So, 'why the sudden uproar?' the undertone rang. A few months later, in June 2024, the call for remigration premiered for the first time in the party's official platform for the European parliamentary elections.[18]

By the time state elections rolled around in Thuringia that September, the AfD's campaign visuals featured a plane soaring through a bright blue sky – a nod to vacation ads of yet another fossil-fuelled industry at the heart of the far right's nostalgia politics, but with a grim twist. The aircraft bore the label 'Deportation Airlines', turning what looked like a cheerful travel ad into a thinly veiled threat. 'Summer, sun, remigration', the slogan read. The party's full embrace of remigration reflects the growing strength of the extreme right wing of the party, which had grown in influence over the past years and once again emerged victorious in internal power struggles. Its electoral base responded favourably, seemingly cheerful and even relieved about the fading of social costs once attached to a politics of organised expulsion so clearly dressed in racist lingo. The AfD soared in the polls even as it let its bourgeois guard down. Finally the people could say 'what they thought all along': 'Germany for the Germans – Foreigners out!' What used to be a far-right slogan shouted at protests had been turned into a song lyric accompanying the beat of the Eurodance hit 'L'amour toujours' across the country. In its most viral case, a young party crowd on Sylt, the affluent North Sea island, filmed itself singing the song in the summer of 2024, with one man pantomiming Hitler's mustache and throwing his right arm up.

To expand its base and push the party system rightward, as it had by the 2024 election, the AfD made immigration – not falling purchasing power or rising costs – the campaign's focal point. The root cause of economic distress and political disillusionment had to be manufactured, which is to say, constructed in discourse and ideology. To that end, the far right capitalised on isolated but high-profile acts of violence by migrants to construct a homogeneous national identity against immigration and to frame all non-Germans as an existential threat.

When an Afghan national injured six people and killed a police officer at a May anti-Islam demonstration in Mannheim, the Junge Alternative staged a memorial event portraying the fallen policeman as a victim of 'failed' immigration policies.[19] After a Syrian asylum seeker killed three people in Solingen three months later, the AfD organised rallies in the city as speakers thundered against 'imported knife crime' and the leader of the far-right wing, Bjorn Höcke, cast the upcoming state election as a choice between him or Solingen. By 'Höcke oder Solingen' he insinuated two options: either ethnonationalist authoritarianism or mass murder. The narrative deepened when, in December 2024, a Saudi-born ex-Muslim drove a car into a crowded Christmas market in Magdeburg, killing six. Sidelining the attacker's history of far-right sympathies and support for the AfD, party leader Alice Weidel seized on the attack as evidence of hatred towards 'us Germans, and us Christians'. At a rally, the crowd erupted in chants of 'Deport them!' when Weidel gestured towards the need for deportations 'so we can finally live once again in security'. After the Magdeburg attack, the share of Germans who viewed migration and asylum as one of the top two political issues rose sharply by 14 points to 37 per cent, making it the most important concern alongside the economy. The series of attacks continued, including a stabbing by an Afghan man with a psychiatric record, and another attack just days before the federal elections in which an asylum seeker drove a Mini Cooper into a trade union rally in Munich, killing a mother and her daughter. The AfD swiftly folded the tragedy into its campaign, declaring once again that 'only we'– and only remigration – 'can stop this migrant violence'. JD Vance, who was in town for the Munich Security Conference when the attack happened, chimed in, 'The real threat to Europe is not Russia, it's not China, it's not any other external actor,' he declared, 'but the threat from within'. 'There is nothing more urgent than mass migration,' Vance concluded.

Retold in succession, these tragedies form a chilling litany. And it is precisely this sequencing that drives the AfD's method of moral-panic politics. By collapsing disparate events into a single, continuous narrative of escalating migrant violence, the far right generated a moral panic accelerating an ongoing process of crisis-driven fascisation. Perpetrators are transformed into an archetype of the dangerous migrant, allowing the far right to inflate specific crimes into evidence of systemic collapse. The AfD curates this pattern carefully, stripping incidents of context, flattening complexity and amplifying racialised fear. Media coverage of the crimes acted as midwife to the frenzy, selectively amplifying or obscuring facts in service of the attention economy. Meanwhile, reporting on the rise in violence against racial minorities following attacks remained largely under the radar. In the AfD's moral panic, migrants become the folk devils of multilayered crisis: they are blamed for draining public resources, eroding cultural identity, social fragmentation and economic decline.

The instrumentalisation of tragedy not only fuelled the AfD's electoral rise but also pushed mainstream parties to adopt tougher immigration stances. Seeing the AfD rising in the polls as the campaign progressed, Merz's initial vow to avoid politicising migration came under strain; it seemed voters were rewarding the AfD's anti-migration focus. The logic of party competition easily eroded any moral scruples or principles the CDU, and in turn other parties, may have had. Using migration and murders as fuel, the AfD drove the entire party system to the right.

Over time, the remigration frame gained traction globally. In September 2024, Austria's far-right FPÖ, long at the forefront of translating identitarian ideology into policy, secured a sweeping electoral victory on a platform centred on mass deportations and anti-Islam hate mongering. The campaign, as the *Guardian* noted, was 'punctuated with the term', with the party even proposing the appointment of an EU-wide 'remigration commissioner'.[20] Momentum quickly spread beyond the Austrian–German axis of the far right. That same month, Donald Trump invoked the term on social media to bolster his campaign, promising to 'return Kamala's illegal migrants to their home countries (also known as remigration)'. The

Spanish far-right party Vox took the remigration rhetoric to the bluetooth speakers of its supporters when it released the song 'Bilette de Vuelta' in February 2025; the accompanying video shows young men of colour in tracksuits waving around knives at blond female flight attendants in dresses and bright lipstick. In April, AfD MP Lena Kotré – Brandenburg's official spokesperson for domestic and migration policy – attended a 'Remigration Conference' hosted by the British fascist Homeland Party, alongside Great Replacement theorist Renaud Camus, where she was welcomed as a politician who 'stands up for the women of her nation and for remigration'. The two issues are 'intrinsically linked', at least in the femonationalist world view of the organisers.[21] At a Lega rally in Florence that same month, Elon Musk took the stage to warn Matteo Salvini that 'mass immigration destroys every country'.

From France to Flanders, Sweden to Slovenia, far-right parties represented in parliament have made remigration official party policy. The global crusade of the remigration doctrine culminated, symbolically, in the May 2025 European Remigration Conference held in Milan. Co-organised by Austrian far-right activist Martin Sellner and framed as a defence of 'free speech', the event brought together around 300 to 400 participants from Europe and the United States in what was described as a blend of identitarian and conservative viewpoints. 'Remigration had a peak victory,' proclaimed co-organiser Alfonso Gozalves of Portugal's Reconquista movement, a group that emerged in 2023 with the stated aim of 'reclaiming' Portugal by expelling those deemed non-Portuguese. While small in number, the remigration summit embodied a rising transnational project. Days after the summit in Italy, Trump's administration announced plans to convert a US State Department bureau historically tasked with aiding the settlement of asylum seekers into an Office of Remigration.

During his feud with California Governor Gavin Newsom amid

the Los Angeles anti-ICE protests in the summer of 2025, Trump ratcheted up his rhetoric: 'The Biden Administration and Governor Newscum flooded America with 21 million illegal aliens, destroying schools, hospitals, and communities, and consuming untold billions of dollars in free welfare,' he wrote on Truth Social, claiming that America had been 'invaded and occupied'. While inflating official estimates of undocumented migrants in the US by nearly 200 per cent, Trump cast himself as the saviour from this deluge: 'I am reversing the invasion,' the post concludes. 'It's called Remigration.' Trump's post was celebrated on Telegram by the mastermind behind the term, Martin Sellner: 'Welcome to Camp Remigration – now officially.'

Manufacturing crisis

As the German far right helped carry remigration from bar room chatter to the corridors of power, Germany's signature export industry was facing a multifaceted crisis of global competition, technological disruption and mounting pressure towards decarbonisation. With the country's economic foundations cracking, the AfD found fertile ground to turn deepening economic insecurity and faltering trust in traditional parties into political mobilisation, offering up migrants and climate policies as culprits of a downturn decades in the making.

Although the AfD has long attacked Germany's energy transition, it was only in 2019 that the party leadership announced that it would henceforth rank climate as the third most important issue in its campaign, with migration and its signature critique of the Eurozone at the top of the agenda. Though climate became an increasingly central theme, the AfD's anti-climate strategy contained contradictions. Notwithstanding the Junge Alternative's pressure to thematise climate change in an eco-fascist light, the main party remained committed to climate denialism, sounded the alarm about blackouts, warned of a looming 'eco-dictatorship' under the Greens, and stoked fears that the rise of climate-conscious politics would flood the country with both migrants and gender-nonconforming people. The themes resonated with its electoral base, but seemingly not beyond: by the time state elections came around in the autumn of 2021 the AfD scored 10.4 per cent of votes, 2.2 per cent less than in the previous elections of 2017.

Jump forward four years and the situation had shifted dramatically: in the parliamentary snap elections of 2025, the AfD won the second most votes with 20.6 per cent of the total, the highest for the far right in Germany since the Second World War. By 2025, climate action had largely disappeared from the electoral agenda entirely, as politicians across parties navigated greenlash. The year prior, far-right and liberal parties and media outlets had joined ranks in constructing the largest moral panic to date about a climate policy – the swap-out of boilers for heat pumps, pejoratively dubbed as the *Heizungshammer* (heating hammer). Having long established itself as the fiercest critic of the Green Party for its seemingly progressive positions on migration, climate action and LGBTQI rights, the AfD marketed itself as the antithesis to the collapsed traffic light coalition – formed by the Social Democratic Party (SPD), the Green Party (Greens) and the Free Democratic Party (FDP), named after their party's colours – and as the only 'alternative' to all establishment parties. The AfD's campaign instrumentalised a sense of crisis and demise that struck a chord with Germans of various political stripes in the lead-up to the elections. And as the traffic light coalition's signal went red, Germany's faltering automobile industry played a crucial role.

A brief history of how the AfD, founded in 2013, leveraged fears surrounding the automotive sector begins with the Volkswagen emissions scandal, or 'Dieselgate'. In 2015 it emerged that Volkswagen had installed 'defeat devices' in diesel vehicles to cheat emissions tests, enabling cars to emit up to forty times the legal limit. The scandal affected millions of vehicles worldwide and led to massive fines, recalls and lawsuits, and a collapse of trust in the German automotive industry as a whole. Moving against the political current, the AfD sought to shield the sector, blaming supposedly ideological pollution targets rather than corporate deception. As German cities began banning older diesel cars to curb toxic air, the party increasingly styled itself as the one that would 'save the diesel'. Stuttgart – both Germany's self-proclaimed car capital and one of its most polluted cities – became a flashpoint in 2019 when the authorities implemented a long-contested ban on older diesel vehicles after years of local activism. The decision triggered months of protests with affected car owners donning yellow vests of their own and declaring, 'We are the people!' The AfD, a minority but conspicuous presence, worked to orchestrate and capitalise on the unrest.[22] Moments like Stuttgart were not isolated but formative staging grounds in the AfD's construction of its carbon populism, in which the 'right to drive a diesel' displaces demands for justice, clean air or accountability for corporate misconduct. From here the party built out its narrative architecture: accusing advocates of transforming the car industry and transport system of 'coercion' and 'delusion', the AfD cast itself as the party of the 'small man', defending his 'freedom' alongside the industry that had made Germany prosperous. While the liberal mainstream sought to revive Germany's image as a climate pioneer, the far right's agitation against the automobile's role in the energy transition was dismissed as a confused culture war – precisely as it was becoming politically consequential.

Already weakened by unstable export markets on the back of the global financial crisis, growing international competition and tightening environmental regulations in the 2010s, Germany's auto industry was dealt a severe blow by the COVID-19 pandemic.

Lockdowns and supply chain breakdowns in 2020 forced factories to halt production, sending output to unprecedented lows. Consumer demand for new vehicles also plummeted amid economic uncertainty, and many smaller auto suppliers did not survive the interruption. While the industry was on a slow path of recovery, Russia's invasion of Ukraine in 2022 and the sharp pivot away from Russian gas and oil unleashed a grave energy crisis: for years cheap Russian gas had been the cornerstone of German industry, and now the taps had been forcibly closed. But the energy crisis triggered by Russia's invasion and Germany's wilful dependence on Russia, maintained despite Putin's authoritarian turn and repeated warnings from Eastern European states, was not the only geo-economic storm that thundered in the pandemic years.[23]

When German car executives visited the Shanghai International Automobile Industry Exhibition for the first time since the onset of the pandemic in 2023, they experienced a 'rude awakening' as China's EV champions unveiled cutting-edge models and battery technology that have since left Western brands scrambling to catch up.[24] The stark rise of Chinese competitors in what used to be a key profit engine for Germany's Big Three (VW, BMW and Mercedes) revealed just how slow German companies had been to recognise the speed of the EV transition.[25]

Amid the turmoil, right-of-centre parties coalesced as defenders of the internal combustion engine, but with different tones: the CDU/CSU calibrated their climate commitments to avoid 'overburdening' manufacturers, while simultaneously lobbying China to slow their EV plans; the FDP championed 'technology-neutral' solutions like e-fuels over bans; and the AfD embraced outright climate scepticism while warning of deindustrialisation. By 2024, the CDU/CSU were openly campaigning against the EU's planned (and later abandoned) 2035 ban on petrol and diesel cars, declaring that 'Germany is a car country' and pledging to preserve

'Germany's cutting-edge combustion engine technology' rather than bow to Brussels's zero-emission mandate.[26] Conveniently for the AfD, long-standing structural weaknesses in the German economy, and in the auto industry in particular, became more apparent after the pandemic, just as the traffic light coalition moved to deliver on its campaign promise to ramp up climate regulations.

Germany's automotive heartland is thus grappling with what one observer called a 'toxic cocktail' of challenges – from pandemic aftershocks and an energy crunch to technological disruption and global competition.[27] By 2024, the cumulative impact of these overlapping crises manifested in cost-cutting decisions that were impossible to hide from the public.[28] Newspapers seemed to feature almost daily headlines warning of factory closures, layoffs and production shifts out of the country. 'Volkswagen is closing three plants in Germany, and Audi is halting production in Belgium, sparking fears of a crisis in Europe's car industry,' ran one such report in October 2024.[29] Major suppliers were also in distress. Bosch, the world's largest car parts maker, revealed it might need to cut 10,000 jobs around the country, blaming weak car demand, high costs and fierce Chinese competition. Soaring energy prices and the EV transition drove many firms to seek cheaper locales. As part of Volkswagen's cost-cutting drive, in December the management announced plans to shift production of its iconic Golf from Wolfsburg to Mexico by 2027, introduce a regressive four-day work week and cut around 35,000 jobs, roughly a quarter of VW's workforce in Germany, by 2030. Labour unions, fearing a domino effect of industrial decline, staged protests and strikes to 'avoid unprecedented plant closures' in Germany.[30]

That the once unstoppable engine of Europe's economy began to sputter provided fortuitous terrain upon which the AfD mobilised. The party's perennial, pre-emptive critique of green industrial policy and its long-standing warnings about 'climate hysteria' and globalist-driven deindustrialisation gained renewed resonance. Blaming the traffic light coalition for job losses and drained wallets, a suffocated industry and factory relocations, the AfD cast itself as the true representative of national industry and the real champion of ordinary people, whom it portrayed as under siege not only from economic decline but also from migration. By linking these two crises under the catch-all solution of national restoration, the AfD echoed the wider international far right in suturing anti-migration politics to a defence of the internal combustion engine. In this context, climate denial – once seemingly unhinged – became politically effective, especially with the Greens in – and on their way out of – government. Parts of AfDs far-right wing even mobilised Zentrum – a self-described 'alternative labour movement' whose founder has documented ties to the terrorist organisation Blood and Honour – to rally Volkswagen and other workers against the EV transition and into support for the AfD. Meanwhile, the party continued to promote a Putin-friendly line, calling to repair and reopen the Nord Stream pipeline, laying the groundwork for an infrastructural alliance between Russian fossil authoritarianism and the European far right. Taken together, the AfD's campaign centred on two pillars of 'returning' to an imagined golden age of racial purity and industrial supremacy: remigration and reindustrialisation.

In the early days of the run-up to the general elections held in February 2024, CDU leader Friedrich Merz declared that migration would not become the central focus of his bid for the chancellorship. Instead, he aimed to shift attention to Germany's stagnating economy. Likewise, none of the contenders emerging from the recently fractured traffic light coalition seemed eager to compete with the increasingly ascendant far right on the topic of migration. Yet by the final days of the campaign, migration had become the dominant issue, with every party but the Left party proposing or endorsing new plans to restrict it. This culminated in what commentators described as the 'falling of the firewall': just weeks before the pivotal federal elections in February 2025, in which the AfD was polling in second place, Merz openly relied on AfD support for the first time to advance legislation tightening migration laws and 'hardening the border' – marking a decisive shift in the overt normalisation of far-right positions within mainstream politics.[31] It was a moment that crystallised the convergence of manufactured crises and crises in manufacturing: as economic insecurity deepened, particularly in Germany's industrial heartlands, the political response had been to displace that anxiety onto migrants – offering a scapegoat where structural solutions were absent.

War on climate

While the remigration discourse has gone global, mainstreaming hyper-racialised narratives and redirecting blame for economic recession onto marginalised populations on its way, war is simultaneously producing a dual effect: it normalises mass death by desensitising publics to large-scale violence, all the while serving as a strategic opportunity for crisis-ridden industries, including the struggling car sector. In this convergence, death and decline are not only obscured but perversely repurposed, turned into tools for nationalist renewal and industrial rejuvenation amid broader systemic and climatic collapse.

War – both on the battlefield and in the political imagination – has become a defining feature of the new phase of climate denial. Following the surge in internationalised civil wars after 2014, the world has seen a rapid escalation in armed conflict: the civil and proxy wars that produced famine in Yemen and Ethiopia; renewed fighting between Armenia and Azerbaijan; escalating violence between India and Pakistan and Pakistan and Afghanistan; Russia's full-scale invasion of Ukraine; Israel's genocide on Gaza which expanded into a regional war including southern Lebanon, Iran and Yemen; and resurging conflicts across the Sahel and Myanmar, to name only a few. From 2021 onward, global levels of armed conflict and battle-related deaths rose sharply each year, culminating in 2024, when the world recorded sixty-one active state-based conflicts across thirty-six countries, the highest number since 1946. The four years following 2021 were the deadliest since the end of the Cold War, with nearly 740,000 people killed. As the US and Israeli war on Iran made clear, this trend is only accelerating, with researchers describing the escalation as not just a peak, but a structural change, driven in part by the internationalisation of war, geopolitical fragmentation and intensifying inter-state competition, both among major and mid-level actors.[32]

War forms the logical antithesis to climate action. It obstructs the international cooperation required for energy transition, while diverting resources – money, metals, manufacturing capacity – into the production of rifles rather than renewables. From an ecological perspective, war nullifies past and present efforts to reduce emissions and protect biodiversity, as meagre as they had previously been. While confidentiality and non-transparency around military infrastructure make it hard to precisely account for its share of greenhouse gas emissions, in a world without armed conflict or war it is estimated to be around 5.5 per cent of the global total.[33] As states rush to become 'war-ready', that number rises. Fossil fuels are the lifeblood of every tank, aircraft and missile built. And then, of course, there is fighting. As Andreas Malm notes, 'Western forces pulverize the living quarters of Palestine by mobilizing the boundless capacity for destruction only fossil fuels can give.'[34] While fossil-fuelled war, wherever it is waged, kills humans and ecosystems in the moment of destruction, it sets in motion long-term cycles of ecological degradation and climate destabilisation – long after the bombs have ceased to fall. A timid extrapolation of greenhouse gas emissions produced in just the first fifteen months of the war in Gaza projected the total to exceed the yearly carbon output of thirty-six countries.[35] In the first three years of the Russian invasion, war-related activities accounted for 237 megatons of total greenhouse gas emissions (MtCO2-eq), roughly equivalent to the combined annual emissions of Austria, Hungary, the Czech Republic and Slovakia.[36] Restoration will itself be a multiplier of emissions. Without even accounting for Trump and Netanyahu's so-called 'Rivieria' construction plan, rebuilding the 100,000 houses bombed within the first two months of the war on Gaza alone is projected to be 'on a par with New Zealand's annual CO_2 emissions and higher than 135 other countries and territories including Sri Lanka, Lebanon and Uruguay'.[37] Every war waged is also a war waged on the climate.

The ongoing militarisation of states around the world can only exacerbate these trends. As rearmament accelerates, investment flows are directed towards and locked into fossil fuel infrastructure. Amidst ongoing Russian aggression, but ultimately triggered by the mounting doubts about continued US support for NATO, Europe is racing to rearm. The European Commission introduced its first ever White Paper on defence in 2025, of which Commission President Ursula von der Leyen, who had previously sold the Green Deal as Europe's 'man on the Moon' moment, was a driving force. The ReArm Initiative foresees that, using an exemption clause, EU member states may raise their defence spending to an estimated €650 billion by 2030.[38] Already in 2024, EU countries collectively spent €326 billion on defence – an increase of 31 per cent compared to 2021. To facilitate a further build-up, in March 2025 the EU temporarily suspended the Maastricht criteria; member states are now permitted to spend up to 1.5 per cent of their GDP on defence for four years, regardless of standard EU debt rules. An added €150 billion is proposed to be generated and channelled into rearmament through joint borrowing.

In Germany that same month, it was the CDU – long-time champions of Germany's constitutional 'debt brake' – that led an equally dramatic reversal of the tradition of fiscal conservatism. CDU leader Friedrich Merz, fresh off an election victory and entering coalition talks with the SPD, pushed through a constitutional amendment exempting all military and security spending above 1 per cent of GDP from borrowing limits. After years of blocking debt brake reform while in opposition, Merz moved swiftly in government, rushing the measure through the outgoing Bundestag before a newly elected chamber – with a strengthened far-right and left – could obstruct the required two-thirds majority. Neither recession-hit Greece nor years of Green Party efforts to unlock climate investment had managed to loosen Germany's strict fiscal rules. In the end, only war could trigger a breakthrough. Repurposing Mario Draghi's famous pledge as ECB president to save the euro as a rallying cry for deficit-funded militarisation, Chancellor Merz declared, 'Whatever it takes.'[39]

Merz's high-spirited commitment to rearmament was no sudden turn, but the culmination of a broader and accelerating trend.[40] In 2024 alone, Germany's military spending surged by 28 per cent to $88.5 billion, catapulting it to the top spot in Europe and making it the fourth-largest military spender globally. Far from being an outlier, Germany joined a sweeping global shift: more than 100 countries ramped up their military budgets in 2024, driving the steepest annual rise in worldwide defence spending since the end of the Cold War. Compared to 2023, global military expenditure jumped by 9.4 per cent. This surge builds on a trajectory that began in 2018, when global military spending began to rise after a period of contraction and flatlining. But far from the excessive spending of 2024 representing a peak, this is the baseline upon which the rewriting of fiscal orthodoxy is taking place. The extraordinary levels of 2024 are now locked in and set to grow.

At the NATO summit in June 2025, the trajectory of militarisation was cemented: nearly all member states agreed to raise their defence spending to 5 per cent of GDP by 2035. NATO Secretary General Mark Rutte hailed the breakthrough, praising Trump's leadership, without which 'we would never, ever, ever have been able to achieve agreement on this'.[41] While sycophantic, his analysis was not entirely wrong: had Trump not strong-armed NATO members by threatening to dismantle the post-war security architecture, Europe's turn to militarisation would likely have been far less pronounced. Spain stood alone in opposing this sweeping redirection of public funds, a position Trump later threatened to punish with higher tariffs.

Trump's aggressive rhetoric towards NATO not only rattled allies but fuelled a speculative boom in European defence equities. When, on the election trail, Trump announced that he 'would encourage' Russia 'to do whatever the hell they want' if others would not increase their spending on NATO, stock value of Rheinmetall AG, one of Europe's largest munitions manufacturers, was catapulted on an uninterrupted climb for over a month, landing a 67 per cent increase.[42] Within days of Trump's re-election, the stock gained another 16 per cent, as traders cashed in on the foreseeable increase in defence spending. Between February 2022 and January 2023 European defence stocks soared by over 300 per cent, while the sharpest growth occurred after mid-2024. By comparison, the STOXX Europe 600 Automobiles and Parts index showed no significant growth over the same period, but opportunities were on the horizon.

While war brings about death and destruction, the car industry sees it as a lifeline. Across the continent, big and small automotive and industrial suppliers are actively seeking partnerships within the defence industry. Rheinmetall is reportedly exploring the conversion of its automotive plants in Berlin and Neuss for weapons manufacturing, already offering workers from a struggling brakes plant jobs in a new munitions factory. One study estimates that current defence spending by European NATO states already supports 680,000 jobs and that raising spending to 3 per cent of GDP would see that number almost double.[43] Rheinmetall alone laid out plans to grow its workforce from 40,000 to 70,000 within two years, providing a home for skilled labourers from the struggling auto industry.[44] As military budgets swell across the continent, rearmament is increasingly framed not only as a geopolitical necessity but as an economic opportunity – particularly for an automobile industry under pressure.

Friedrich Merz's government all but flaunts the idea that rearmament should jump-start the engines of an ailing car industry. For example, the government treaty reads,

> The steel and automotive industries are facing enormous structural challenges. At the same time, the defense industry must scale up rapidly and on a large scale. We will therefore examine how the conversion and upgrading of existing facilities can be supported to meet the needs of the defense industry.[45]

Yet, of course, scale remains an issue: as Germany's auto sector generated more than €540 billion in revenue in 2024, compared to just €30 billion across its five largest arms manufacturers the year prior, the defence sector will not offer full salvation to a crisis-ridden car industry. But the fact that it can absorb excess capacity, safeguard jobs and redirect industrial infrastructure gives political actors and industry leaders alike a strong incentive to support continued rearmament. Rearmament, in this light, is no longer just about defence; it is becoming a lifeline to save a sputtering industrial base. Or as Deutsche Bank optimistically notes, amid global instability 'Germany has a historic opportunity to kill two birds with one stone by turning some of its auto-making prowess to military production.'[46] In this discourse, no words are lost on one of the root causes that brought the car industry to its knees: the mandate to decarbonise and its associated pressures. As car manufacturers preside over massive manufacturing capacities but demand dwindles and profit margins shrink, parts of the infrastructure for mass-producing cars is repurposed to mass-produce tanks.

As larger chunks of GDP are directed towards the military, the expectation grows that such investments can not only revive a sluggish auto sector, but also – as the *Wall Street Journal* puts it – 'unlock new engines for growth and exports' across the economy. One report entitled *Guns and Growth* anticipated that increased defence spending would be a shot in the arm for European economies, swapping US imports for domestically produced weapons.[47] Keir Starmer echoed this logic, promising that the increase in defence spending – first from 2.3 to 2.5 per cent of GDP, and later to 5 per cent – would 'put more money in people's pockets' and make 'communities across the UK better off'. Starmer's words mark a shift towards military Keynesianism, where the promise of economic revival through the machinery of war – not the ecological transition – is framed as an engine to create 'good, skilled jobs'.[48]

Beyond the battlefield, mass death becomes institutionalised as cost-cutting policy. In the name of putting America First 'one dollar at a time', the dismantling of USAID marked one of the most sweeping contractions of humanitarian assistance in modern history. While USAID's programmes have long been entangled with US geopolitical interests, its dismantling nonetheless represented a profound blow to global humanitarian infrastructure. By that time – even as Musk, who had been tasked to 'root out waste' and 'block woke programs' was being congratulated by Trump for 'slashing the fat' after five months of work – an estimated 300,000 people had died as a result of the cuts, more than 200,000 of them children.[49]

By 2030 some 14 million people could die as a result of the cuts, a *Lancet* study projected.[50] Even a twelve-month phase-out might have allowed for contingency planning and saved countless lives. Importing the Silicon Valley mantra of 'move fast and break things' into government, speed became the currency of moral repair and monetary savings: the American taxpayer, wounded by an apparent lack of reciprocal 'generosity', and the fiscal budget, bloated by the misuse of aid, could be reconciled by rendering the lives of supposedly ungrateful and undeserving foreigners expendable.[51]

In this emerging regime of militarised accumulation and a redefinition of global cooperation, death is not merely a consequence of crisis, but a condition for economic revival and budgetary discipline. The pandemic and right-wing 'freedom' movements, premised on opening the economy no matter the health risks and death tolls, arguably already ingrained this necropolitical logic into popular consciousness, but increasing militarisation and the ascendance of far-right neoliberalism mainstream it further still. War, displacement and abandonment are rendered productive – transformed into the raw material upon which a stagnant economy is retooled and a collapsing social order is maintained.

As a near consensus among leaders is hardening around rearmament and making nations 'war-ready', the far right is positioning itself as a force of peace in Europe. Figures on the far right cultivate ties to Putin, casting him as the archetype of strong, sovereign leadership in contrast to what they frame as the weakness and hypocrisy of Western democracies. This alignment is no accident: authoritarianism, fossil fuel dependency and climate denial are tightly interwoven in the far-right project. By rejecting both military aid to Ukraine and climate action as elite-driven and destabilising ventures, the far right exploits war fatigue and economic anxiety to present itself as the only political force promising energy security and national stability. Likewise, as the bombs dropped on Gaza, parts of the far right positioned themselves as a bulwark against anti-Semitism.

Crusaders for 'peace'

As part of a broader strategy to defend and mainstream far-right ideology, many parties have sought to deflect accusations of anti-Semitism by aligning themselves with the state of Israel and by reaching out to Jewish communities, emphasising a supposedly shared Judaeo-Christian heritage. Already in 2014, Marine Le Pen made a direct appeal to Jewish voters, claiming that the Front national was 'without a doubt the best shield to protect you against the one true enemy, Islamic fundamentalism'. That her father and party co-founder, Jean-Marie Le Pen, had previously been convicted for downplaying the Holocaust was apparently no contradiction. From conservative and far-right coalitions in the EU parliament to Javier Milei and Jair Bolsonaro in South America, the strategy of superficially championing anti-anti-Semitism helped unite the international far right during the most acute phase of the genocide. In May 2025, Netanyahu and the state of Israel invited far-right leaders from across Europe to a conference in Jerusalem on Combating Antisemitism, with French National Rally party leader Jordan Bardella delivering a keynote speech blaming anti-Semitism on migration. 'Islamism is the totalitarianism of the 21st century' that threatens 'to destroy everything that is not like it', he claimed amidst a genocide in Gaza.[52] In planning to honour Charlie Kirk with a posthumous award at the 2026 Combatting Antisemitism conference, organisers promised a more 'mainstream' event, which featured far-right leaders Santiago Abascal of Spain, Jimmie Åkesson of Sweden and Eduardo Bolsonaro of Brazil, as well as panels on 'Denial as a Weapon' and 'Antisemitism and Radicalisation in the West'.

But beyond the superficial benefits of an anti-anti-Semitic facelift, the European right also harbours a deeper ideological admiration for Israel as an ethnonationalist model: a state seen to embody rigid border control and unapologetic nationalism. In this projection, the complexities of Israeli society and the lived realities of Palestinians are erased, replaced by a one-dimensional fantasy tailor-made to serve the needs of right-wing identity politics in Europe. As the genocide in Gaza produced international pressure on the Israeli state – from protests in the streets to arrest warrants for the Israeli leadership at the ICC – the right seized the moment to intensify its calls for 'ending illegal mass migration' while mainstream conservatives like Merz echoed such calls by vowing to end 'imported anti-Semitism'.

This strategic repositioning of the far right in Europe finds a minor parallel in the shifting political landscape of the United States, where Israel's war on Gaza catalysed a rupture within another historically loyal voting bloc: Muslim and Arab-American communities. Nationally, the majority of these communities backed Biden in 2020. But in 2024, in response to Biden's unwavering support for Israel's war on Gaza, many voters broke ranks. Nowhere was this more visible than in Dearborn, Michigan, which is home to a large Muslim population drawn historically by jobs in the auto industry and which historically votes Democratic. In 2020 Joe Biden carried the city with nearly 70 per cent of the vote. On the back of the war in Gaza, many in Dearborn and surrounding cities turned to Trump, despite his history of anti-Muslim policies like the travel ban. The result: Trump took Dearborn, beating Harris by 6 percentage points. Perversely, the far right can benefit from ongoing wars that fracture traditional political alliances, simultaneously claiming to be anti-war while glorifying genocidal violence. Such local lessons are reflected in global inversions, with a prime example in Trump's 'Board of Peace': a body first created to control the fate of Gaza as a so-called riviera of privatised real estate, and then imperialistically pitched as a privatised UN with billion-dollar membership fees and Trump as its permanent, all-powerful Secretary General.

While backing the genocide in Gaza, the US government initiated a series of kidnappings targeting immigrants who spoke against it. With the illegal detention of Mahmoud Khalil, the campaign against protesters set the stage for a brutal spectacle of 'mass deportation' in which masked ICE agents disappeared migrants from one city to the next. While expanding his first-term 'travel ban' on Muslim-majority countries, the Trump administration also effectively ended the right to asylum – dropping the cap on admissions from 125,000 to 7,500 – with the notable 'humanitarian' exception of white Afrikaner farmers, many of whom were granted refugee status. 'There is a genocide happening,' Trump claimed at a press conference shortly after fifty-nine Afrikaners arrived in a US government-chartered plane. The upside-down deflection echoed Elon Musk, who had previously accused the South African government of enacting 'racist ownership laws' and committing a 'genocide against white people'. Of course, the distinction between welcomed guest and scorned invader also tracks the racialised contours of class. Accentuating the same divide from an inverse angle, Trump directed the Department of Justice to establish a system for selling 'gold card' immigrant visas, priced at around $5 million each. The dynamics are not unprecedented as the US has long opened the door for the wealthy while policing the poor. What sets the Trump era apart is the violent enforcement and unapologetic visibility it gives to these hierarchies while overturning the progressive achievements of a period it labels 'reverse discrimination'.

The case of Naomi Seibt exemplifies how the far right stretches this logic from wealth into ideology. Once branded the 'anti-Greta' for her climate denialism, the far right influencer with close ties to the climate denialist Heartland Institute, the AfD and the MAGA movement applied for asylum in the United States in November 2025, claiming persecution for her political views. Her appeal coincided with reports and proposals that the Trump administration was prioritising 'free speech advocates in Europe' and other applicants 'targeted for peaceful expression of views online such as opposition to mass migration or support for populist political parties'.[53] As the inversion of persecution and asylum advances, a system set in place to safeguard those fleeing war or repression is retooled into an instrument of ideological and racial alignment.

This realignment reflects the contours of what Ernst Fraekel, writing about National Socialism in 1941, coined the dual state: one in which parallel legal and administrative systems operate according to entirely different logics.[54] On one side, a transactional, privilege-based system facilitates mobility and protection for the wealthy, white and politically aligned. On the other, a punitive apparatus targets the racialised poor through surveillance, detention and deportation. Both are part of the same regime: unequal by design, but increasingly executed in coordination. The Trump administration's immigration policy lays bare how these dual logics no longer compete behind the scenes; they function in tandem, reinforcing a hierarchy that is as old as the republic, but stripped of pretence and channelled into a fossil fascist project.

As climate breakdown accelerates, these dynamics are more likely to accelerate than to recede. The infrastructures being built today – by agencies like ICE and companies like Palantir – are not simply built to enforce harsher borders; they are designed to automate and expand them, preparing a future in which mobility itself becomes a privilege of the few. In this 'regime change from the right', surveillance and biometric control will govern who gets to flee disaster, who is abandoned to die in it, and who profits from the management of that death. Colombian President Gustavo Petro's warning rings with chilling clarity: 'What we see in Gaza is the rehearsal of the future.'

Of lifeboats and yachts

In the context of the climate crisis, mobility politics are inflected by a carbon populism in which the right of 'the people' to move and emit – via SUVs, low-cost flights or long commutes – is treated as sacrosanct. Fossilised populations, like fossil capital itself, must keep moving, all while the mobility of racialised and displaced Others is increasingly restricted. This tension is compounded by intensifying geopolitical rivalries, where competing elites construct parallel energy infrastructures and mobility regimes, one brown and fossil-fuelled, the other green and supposedly 'clean'. Militarisation fuels simultaneously the West's defence-industrial base and a securitisation of its border regimes. In this landscape, mobility becomes a sorting mechanism – who gets to move freely, and who gets contained – structured by carbon privilege and enforced by military power.[55]

Having traced the contemporary politics of mobility across three intersecting spheres – climate, migration and militarisation – we can see how the trajectory of Western governments increasingly resembles a fascistic environmental moral philosophy. Fifty years after Garrett Hardin published his 'Lifeboat Ethics' essay – an infamous tract in economics centred on mobility – environmental racism is developing into a historical force. Writing against the backdrop of an environmental movement propelled by the publication of *The Limits to Growth*, Hardin reacted to the idea of 'Spaceship Earth', in which all passengers travel through space jointly in a vehicle with finite resources. To Hardin, the metaphor invited 'misguided idealists' to 'justify suicidal policies for sharing our resources through uncontrolled immigration and foreign aid'. The outcome would be 'complete justice, complete catastrophe', he concluded. 'The boat swamps, everyone drowns.'[56]

Hardin's alternative metaphor of the lifeboat shifted the burden of the 'ethical' responsibility in dealing with finite resources onto the global South. In typical neo-Malthusian fashion, Hardin argued that the root problem of resource strains lies with the 'reproductive differences between the rich nations and the poor nations'. In his reading, migrants and recipients of aid were rocking the boat, not a capitalist world system that unevenly sequesters resources. But it is one line of this gruesome essay that carries particular relevance for the new phase of denial we have sketched in this book. Hardin lamented that 'a true spaceship would have to be under the control of a captain, since no ship could possibly survive if its course were determined by a committee'. Shunning the United Nations as a 'toothless tiger', Hardin's rant must be understood as an assault on the very idea of democracy. He instead prescribed a singular 'captain' who can 'enforce' lifeboat ethics upon everyone else. Such neo-Malthusian prescriptions resonate as authoritarian leaders defend models of fossil-fuelled growth and mass deportations, while slashing social welfare, international aid and climate subsidies.

Trump is, of course, prototypical of Hardin's fascist captain. But many far-right leaders are casting themselves in this mould. Together they decry modest climate action as an apocalypse and push for a policy of 'remigration' executed by an expansive militarised apparatus. While dismissing the climate crisis, the captain points at the sinking boat in the sea. This despite the fact that the real culprits of climate breakdown come in a private yacht, not an overloaded lifeboat.[57]

Acknowledgements

We are grateful to Rosie Warren, Richard Seymour, the entire Salvage Collective and Leo Hollis at Verso for their early encouragement to expand this project from an essay into a book – and for their support throughout its completion. For their comradely feedback and support as we expanded and deepened our analysis, we would like to thank Alyssa Battistoni, Lise Benoist, Marius Bickhardt, Anoushka Zoob Carter, Valentin Domann, Michel Feher, Ståle Holgersen, Andreas Malm, John Munro, Dieter Plehwe, Quinn Slobodian, Michael Söding, Oscar Talbot, Troy Vettese, as well as audiences at the Historical Materialism conference in London, the Petrocultures conference at the University of Southern California, the Conservative Critiques conference at Princeton University, the Subversive Forum in Zagreb, and the Political Ecologies of the Far Right conference at Uppsala University. Thanks also to Svenska Forskningsrådet Formas (2018-01702) for the support of the White Skin, Black Fuel grant.

Notes

Introduction

1 Andreas Malm and the Zetkin Collective, *White Skin, Black Fuel: On the Danger of Fossil Fascism* (Verso, 2021), 3–4.
2 Ibid.
3 Stuart Hall, Chas Critcher, Tony Jefferson, John Clarke and Brian Roberts, *Policing the Crisis: Mugging, the State, and Law and Order* (Macmillan Press, 1978), 323.
4 Peter J. Jacques, 'A General Theory of Climate Denial', *Global Environmental Politics*, 12.2 (2012). See also Tad Delay, *Future of Denial: The Ideologies of Climate Change* (Verso, 2024); Kristoffer Ekberg, Bernhard Forchtner, Martin Hultman, Kirsti M. Jylhä, *Climate Obstruction How Denial, Delay and Inaction Are Heating the Planet* (Routledge, 2023).
5 See the *War on Cars* podcast; for their book, see Sarah Goodyear, Doug Gordon and Aaron Naparstek, *Life after Cars: Freeing Ourselves from the Tyranny of the Automobile* (Thesis, 2025).
6 Matthew Paterson, *Automobile Politics: Ecology and Cultural Political Economy* (Cambridge University Press, 2007).
7 This definition builds on the 'extractive populism' of Shane Gunster et al. See Shane Gunster, Robert Neubauer, John Bermingham and Alicia Massie, '"Our Oil": Extractive Populism in Canadian Social Media', in William K. Carroll, ed., *Regime of Obstruction: How Corporate Power Blocks Energy Democracy* (Athabasca University Press, 2021).
8 William Davies, 'TV Meets Fruit Machine', *London Review of Books*, 47.11 (2025), 4–5.

9 William Callison and Verónica Gago, 'The Chainsaw International', *Boston Review*, 3 April 2025.
10 James J. Patterson, 'Backlash to Climate Policy', *Global Environmental Politics*, 23.1 (2023).
11 Andreas Folkers, 'Fossil Modernity: The Materiality of Acceleration, Slow Violence, and Ecological Futures', *Time and Society*, 30.2 (2021).
12 International Energy Agency, *The Oil and Gas Industry in Net Zero Transitions* (2024), 26.
13 Simon Pirani, *Burning Up: A Global History of Fossil Fuel Consumption* (Pluto Press, 2018).
14 Thea N. Riofrancos, *Extraction: The Frontiers of Green Capitalism* (W. W. Norton & Company, 2025); Paris Marx, *Road to Nowhere: What Silicon Valley Gets Wrong about the Future of Transportation* (Verso, 2022).
15 Jean-Baptise Fressoz, *More and More and More* (Penguin, 2024).
16 Alexandre Milovanoff, I. Daniel Posen and Heather L. MacLean, 'Electrification of Light-Duty Vehicle Fleet Alone Will Not Meet Mitigation Targets', *Nature Climate Change*, 10.12 (2020).
17 Michael J. Albert, 'Ecosocialism for Realists: Transitions, Trade-Offs, and Authoritarian Dangers', *Capitalism Nature Socialism*, 34.1 (2023).

1. The Inverted Crisis

1 Stuart Hall, 'The Great Moving Right Show', in *Selected Political Writings* (Duke University Press, 2017), 177.
2 Ibid., 33.
3 See for example Alec Luhn, 'Sooner-Than-Expected Climate Impacts Could Cost the World Trillions', *New Scientist* (14 January 2026); 'Global Economy Could Face 50% Loss in GDP Between 2070 and 2090 from Climate Shocks, Say Actuaries', *Guardian* (16 January 2025); Kamiar Mohaddes and Mehdi Raissi, 'Rising Temperatures, Melting Incomes: Country-Specific Macroeconomic Effects of Climate Scenarios', *PLOS Climate* 4.9 (2025); Maximilian Kotz, Anders Levermann, Leonie Wenz, 'Author Correction of "The Economic Commitment of Climate Change"' (6 August 2025), zenodo.org/records/15984134.
4 Geoff Mann, 'Treading Thin Air', *London Review of Books*, 45.17 (2023). Ståle Holgersen, *Against the Crisis: Economy and Ecology in a Warming World* (Verso, 2024), 91.
5 Sarah Schöngart, Zebedee Nicholls, Roman Hoffmann, Setu Pelz and

Carl-Friedrich Schleussner, 'High-Income Groups Disproportionately Contribute to Climate Extremes Worldwide', *Nature Climate Change*, 15 (2025).

6 One study which foregrounds the political-ecological sensitivities within Gramsci's writings is Michael Ekers, Gillian Hart, Stefan Kipfer and Alex Loftus, *Gramsci: Space, Nature, Politics* (John Wiley & Sons, 2012).

7 Holgersen, *Against the Crisis*. Holgersen is also a member of the Zetkin Collective.

8 Holgersen, *Against the Crisis*, 89–91. See also Ståle Holgersen, 'What Are We Waiting For?', *Spectre*, 5 August 2025.

9 See James Meadway, 'The Minsky Moment', *Macrodose*, 12 October 2025.

10 Building on the work of Claus Offe, the dilemma of climate-related insurance was recently underscored by Stephen J. Collier, 'Insurance and the "Irrationalization" of Disaster Policy: A Political Crisis Theory for an Age of Climate Risk', *British Journal of Sociology* (2025).

11 Adam Tooze, *Crashed: How a Decade of Financial Crises Changed the World* (Penguin, 2018); Daniela Gabor, 'The Wall Street Consensus', *Development and Change*, 52.3 (2021).

12 This paragraph follows an argument set out in Ulrich Brand et al., 'Structural Limitations of the Decarbonization State', *Nature Climate Change*, 15.9 (2025).

13 Holgersen, *Against the Crisis*, 166.

14 Antonio Gramsci, 'The International Situation and the Struggle against Fascism, 1926', in David Beetham, ed., *Marxists in the Face of Fascism* (Haymarket, 2019), 125.

15 Antonio Gramsci, *Selections from the Prison Notebooks* (Lawrence & Wishart, 1971), 321.

16 Andrew Sayer, 'Moral Economy as Critique', *New Political Economy*, 12.2 (2007); Andrew Sayer, 'Moral Economy and Political Economy', *Studies in Political Economy*, 61.1 (2000). Jeremy Adelman, 'Introduction. The Moral Economy: The Careers of a Concept', *Humanity: An International Journal of Human Rights, Humanitarianism, and Development*, 11.2 (2020).

17 Yasmin Afshar, 'Psycho-soziale Aspekte der Automobilität heute', in Psychologists/Psychotherapists for Future e.V., eds, *Kritische Umweltpsychologie* (Psychosozial-Verlag, 2024). Tadzio Mueller, *Zwischen friedlicher Sabotage und Kollaps* (Mandelbaum, 2024). Geoff Pfeifer, 'Of Automobility and Authoritarian Desire: Althusser, Deleuze and Guattari, and Authoritarian Subjectivity in Climate Politics and Beyond', forthcoming.

18 Stanley Cohen, *Folk Devils and Moral Panics: Creation of Mods and Rockers* (MacGibbon and Kee, 1972), 1.

19 Stuart Hall, Chas Critcher, Tony Jefferson, John Clarke and Brian Roberts, *Policing the Crisis: Mugging, the State, and Law and Order* (Macmillan Press, 1978).

20 Ibid., 322.

21 Ibid., 219. See Michael Feola, *The Rage of Replacement* (University of Minnesota Press, 2024); Cynthia Miller-Idris, *Man Up: The New Misogyny and the Rise of Violent Extremism* (Princeton University Press, 2025).

22 See Judith Butler, *Who's Afraid of Gender?* (Verso, 2024); and Jules Gill-Peterson, *A Short History of Trans Misogyny* (Verso, 2024).

23 See Adrian Daub, *The Cancel Culture Panic: How an American Obsession Went Global* (Stanford University Press, 2024). William Callison and Verónica Gago, 'The Far-Right International Versus Anti-Fascist Internationalism', *South Atlantic Quarterly* 1, 124.4, 2025.

24 On such tipping points in Germany, see Daniel Mullis, Maximilian Pichl, and Vanessa E. Thompson. 'Am autoritären Kipppunkt', *taz.de* (16 June 2023).

25 Robert O. Paxton, *The Anatomy of Fascism* (Penguin, 2007), 218.

26 Naomi Klein, *Doppelganger* (Penguin, 2024), 156.

27 Ibid., 156.

28 See Hannah Arendt, *The Origins of Totalitarianism* (Harcourt Books, 1976), Chapter 13.

29 Wendy Brown, *In the Ruins of Neoliberalism: The Rise of Antidemocratic Politics in the West* (Columbia University Press, 2019); Wendy Brown, *Nihilistic Times: Thinking with Max Weber* (Harvard University Press, 2023).

30 William Callison and Quinn Slobodian, 'Coronapolitics from the Reichstag to the Capitol,' *Boston Review*, 12 January 2021; Sam Moore and Alex Roberts, *Post-Internet Far Right* (Dog Section Press, 2021); Carolin Amlinger and Oliver Nachtwey, *Gekränkte Freiheit Aspekte des Libertären Autoritarismus* (Suhrkamp, 2022); Aris Komporozos-Athanasiou, *Speculative Communities: Living with Uncertainty in a Financialized World* (University of Chicago Press, 2022); Derek Beres, Matthew Remski and Julian Walker, *Conspirituality: How New Age Conspiracy Theories Became a Health Threat* (Random House, 2023).

31 Gareth Watkins, 'AI: The New Aesthetics of Fascism', *New Socialist*, 9 February 2025.

32 'Half of UK Adults Worry That AI Will Take or Alter Their Job, Poll Finds', *Guardian*, 27 August 2025.

33 Naomi Klein and Astra Taylor, 'The Rise of End Times Fascism', *Guardian*, 13 April 2025.

34 Edmund Neill, *Conservatism* (John Wiley & Sons, 2021), 18; see also Matthew

McManus, *The Political Right and Equality: Turning Back the Tide of Egalitarian Modernity* (Routledge, 2024).

35 Neill, *Conservatism*, 14.

36 Michael Freeden, *Ideologies and Political Theory: A Conceptual Approach* (Clarendon Press, 1998), 336.

37 Corey Robin, *The Reactionary Mind: Conservatism from Edmund Burke to Donald Trump* (Oxford University Press, 2017), 40, 47.

38 Klein, *Doppelganger*, 116.

39 Jen Schneider, Steve Schwarze, Peter K. Bsumek and Jennifer Peeples, *Under Pressure: Coal Industry Rhetoric and Neoliberalism* (Palgrave Macmillan, 2016). A similar dynamic has been observed in the Canadian oil and gas industry's efforts to cultivate a countermovement by equating 'existential threats' to the industry with threats to ways of life, a discourse which 'inverts the common framing of the climate crisis'. See Jordan Kinder, *Petroturfing: Refining Canadian Oil Through Social Media* (University of Minnesota Press, 2024), 213.

40 Ibid., 25.

41 Ibid., 33, 45.

42 Ibid., 15.

43 Antonio Gramsci, 'The Ape People', *L'ordine Nuovo*, 2 January 1921.

44 Ibid.

45 Paxton, *The Anatomy of Fascism*, 218.

46 Roger Griffin, *Fascism* (Polity, 2018), 46.

47 Geoff Eley, 'What Is Fascism and Where Does It Come From?', *History Workshop Journal*, 91.1 (2021).

48 Roger Griffin, 'The Primacy of Culture: The Current Growth (or Manufacture) of Consensus Within Fascist Studies', *Journal of Contemporary History*, 37.1 (2002).

49 Ibid., 37.

50 Paxton, *The Anatomy of Fascism*, 218.

2. The Motor of History

1 Karl Marx, Capital: *A Critique of Political Economy* (Penguin, 1990).

2 Antonio Gramsci, *Selected Prison Notebooks* (Lawrence and Wishart, 1971), 302.

3 Ibid., 298; Nicole Shukin, *Animal Capital: Rendering Life in Biopolitical Times* (University of Minnesota Press, 2009), 159.

4 Henry Ford and Samuel Crowther, *My Life and Work* (CruGuru, 2008), 34.

5 James P. Womack, Daniel T. Jones and Daniel Roos, *The Machine That Changed the World* (Collier Macmillan, 1990).

6 Ford and Crowther, *My Life and Work*, 115.

7 M. E. O'Brien, *Family Abolition: Capitalism and the Communizing of Care* (Pluto Press, 2023), 117.

8 Michel Aglietti, *A Theory of Capitalist Regulation* (Verso, 1979), 152.

9 Quoted in Winfred Wolff, *Car Mania* (Pluto Press, 1996), 72.

10 Ford and Crowther, *My Life and Work*, 6: 'It is the function of business to produce for consumption and not for money or speculation.'

11 Ford owned coal and silica mines, iron ore deposits, forests and even a rubber plantation and workers' colony in the Amazon called Fordlandia. His railways and cargo ships delivered resources to be transformed on site at one of his facilities.

12 Womack et al., *The Machine That Changed the World*, 39.

13 Anton Jäger, 'Populists, Producers and the Politics of Rentiership', *Political Quarterly*, 91.2 (2020); Stefan Link, 'Rethinking the Ford–Nazi Connection', *Bulletin of the GHI*, 49 (2011).

14 Hasia Diner, *Ford's Anti-Semitism*, PBS.

15 This was the region where Ford himself was raised. John Abromeit, *Transformations of Populism in Europe and the Americas: History and Recent Tendencies* (Bloomsbury Academic, 2016).

16 Ford and Crowther, *My Life and Work*, 58.

17 Michel Feher, *Producteurs et parasites: L'imaginaire si désirable du Rassemblement national* (La Découverte, 2024). Daniel HoSang and Joseph E. Lowndes, *Producers, Parasites, Patriots: Race and the New Right-Wing Politics of Precarity* (University of Minnesota Press, 2019).

18 Jäger, 'Populists, Producers and the Politics of Rentiership'.

19 Stefan Link, *Forging Global Fordism: Nazi Germany, Soviet Russia, and the Contest over the Industrial Order* (Princeton University Press, 2023), 3.

20 Link, 'Rethinking the Ford–Nazi Connection'.

21 At the late-1930s exchange rate, 1,000 reichsmarks roughly equalled $300, which if adjusted into purchasing power today would be approximately $11,000.

22 'Volkswagen', in *Holocaust Encyclopedia*, at encyclopedia.ushmm.org.

23 Adam Tooze, *The Wages of Destruction: The Making and Breaking of the Nazi Economy* (Penguin, 2007), 185.

24 Massimo Moraglio, *Driving Modernity: Technology, Experts, Politics, and Fascist Motorways, 1922–1943* (Berghahn Books, 2017).

25 David Matless, *Landscape and Englishness*, 2nd expanded edition (Reaktion

Books, 2016); Mick Hamer, *Wheels Within Wheels: A Study of the Road Lobby* (Routledge & Kegan Paul, 1987).

26 Link, *Forging Global Fordism*, 3.

27 Ibid.

28 Kat Eschner, 'How Detroit Went from Motor City to the Arsenal of Democracy', *Smithsonian Magazine*, 28 March 2017.

29 Peter F. Drucker, *Concept of the Corporation* (Transaction, 2008), 176.

30 This concept was pitched by Marx and later popularised by John Bellamy Foster.

31 Éric Pineault, *A Social Ecology of Capital* (Pluto Press, 2023).

32 The 'Anthropocene's battering ram' is historian Bryan Appleyard's coinage. See *War on Cars* podcast, 'Traffication with Paul Donald' (18 July 2023), episode 108. See also Bryan Appleyard, *The Car: The Rise and Fall of the Machine That Made the Modern World* (Weidenfeld & Nicolson, 2022).

33 Daniel Yergin, *The Prize: The Epic Quest for Oil, Money and Power* (Simon & Schuster, 1991), 62–3.

34 Timothy C. Winegard, *The First World Oil War* (University of Toronto Press, 2016), 5.

35 Yergin, *The Prize*, 179.

36 Timothy Mitchell, *Carbon Democracy: Political Power in the Age of Oil* (Verso, 2013), 204.

37 Christophe Bonneuil and Jean-Baptiste Fressoz, *The Shock of the Anthropocene: The Earth, History and Us* (Verso, 2016), 123.

38 Drucker, *Concept of the Corporation*, 234.

39 Bonneuil and Fressoz, *The Shock of the Anthropocene*, 129–31.

40 Ibid., 135–8.

41 Adam Hanieh, *Crude Capitalism: Oil, Corporate Power and the Making of the World Market* (Verso, 2024), 46.

42 Paterson, *Automobile Politics*.

43 Eric Hobsbawm, *The Age of Extremes: The Short Twentieth Century, 1914–1991* (Abacus, 2011).

44 Aglietti, *A Theory*, 159; Matt Huber, *Lifeblood: Oil, Freedom and the Forces of Capital* (University of Minnesota Press, 2013).

45 David Harvey, *Rebel Cities: From the Right to the City to the Urban Revolution* (Verso, 2019), 89. By 1957, two thirds of Americans were in debt.

46 Malm and Zetkin, *White Skin, Black Fuel*, 576; Jason Henderson, 'Secessionist Automobility: Racism, Anti-urbanism, and the Politics of Automobility in Atlanta, Georgia', *International Journal of Urban and Regional Research*, 30.2 (2006).

47 Stuart Hall, 'The Supply of Demand', in *Selected Political Writings* (Duke University Press, 2017), 53.

48 Paterson, *Automobile Politics*.

49 Andre Gorz, 'The Social Ideology of the Motorcar', *Libcom*, 2017.

50 Simon Pirani, *Burning Up: A Global History of Fossil Fuel Consumption* (Pluto Press, 2018), 84; Elizabeth Chatterjee, 'Towards an Energetics of Class: Comparing Energy Protests in India and the United States', *Comparative Studies in Society and History*, 66.3 (2024).

51 Paterson, *Automobile Politics*; Hamer, *Wheels Within Wheels*.

52 Giulio Mattioli, Cameron Roberts, Julia K. Steinberger and Andrew Brown, 'The Political Economy of Car Dependence: A Systems of Provision Approach', *Energy Research and Social Science*, 66 (2020), 14.

53 Neil Davidson, *Nation-States: Consciousness and Competition* (Haymarket Books, 2016), 174.

54 Mattioli et al., 'The Political Economy of Car Dependence', 4 (emphasis added).

55 Quoted in Davidson, *Nation-States*, 236 (emphasis added).

56 Mike Davis, 'Marx's Lost Theory', *New Left Review*, 93 (2015).

57 Tim Edensor, *National Identity, Popular Culture and Everyday Life* (Routledge, 2020), 123.

58 Ibid., 122.

59 David Edgerton, 'The Contradictions of Techno-nationalism and Techno-globalism: A Historical Perspective', *New Global Studies*, 1.1 (2007).

60 Richard Nelsson, 'Cars from the Continent: The Direction of Travel in 1960 – from the Archive', *Guardian*, 17 June 2020.

61 Edensor, *National Identity*, 123.

62 Womack et al., *The Machine That Changed the World*, 47.

63 Edensor, *National Identity*, 123.

64 Ibid.

65 Dominic Sandbrook, *Never Had It So Good: A History of Britain from Suez to the Beatles* (Abacus, 2008), 121.

66 Ibid.; Manu Goswami, 'Rethinking the Modular Nation Form: Toward a Sociohistorical Conception of Nationalism', *Comparative Studies in Society and History*, 44.4 (2002).

67 Ibid., 789.

68 Hobsbawm, *The Age of Extremes*, 273.

69 Simon Clarke, 'The Global Accumulation of Capital and the Periodisation of the Capitalist State Form', in Werner Bonefeld, Richard Gunn and Kosmas Psychopedis, eds, *Open Marxism*, vol. 1 (Pluto Press, 1992), 3.

70 Theodore Adorno, *Philosophical Elements of a Theory of Society* (Polity Press, 2019), 38.

71 Hanieh, *Carbon Capitalism.*

72 Bonneuil and Fressoz, *The Shock of the Anthropocene*, 164.

73 Endnotes Collective, 'A History of Separation', in *Unity in Separation* (Endnotes, 2015).

74 Kristin Ross, *Fast Cars, Clean Bodies: Decolonization and the Reordering of French Culture* (MIT Press, 1995), 16.

75 David Edgerton, *The Rise and Fall of the British Nation: A Twentieth-Century History* (Penguin, 2019), 283.

76 Planka.nu, *Traffic Power Structure* (PM Press, 2016), 27.

77 Hobsbawm, *The Age of Extremes*, 264.

78 Bonneuil and Fressoz, *The Shock of the Anthropocene*, 165.

79 Adam Hanieh and Rafeef Ziadah, 'Misperceptions of the Border: Migration, Race, and Class Today', *Historical Materialism*, 31.3 (2023).

80 Karl Marx, *Capital: A Critique of Political Economy* (Penguin, 1990), 275.

81 Étienne Balibar, 'Communism and Citizenship' in *Equaliberty: Political Essays* (Duke University Press, 2014), 160.

82 Wolfgang Streeck, 'Industrial Relations and Industrial Change: The Restructuring of the World Automobile Industry in the 1970s and 1980s', *Economic and Industrial Democracy*, 8.4 (1987), 438.

83 Wolfgang Streeck, 'Citizens as Customers', *New Left Review*, 76 (2012).

84 For an overview of this misleading framing of the oil crisis, see Huber, *Lifeblood*; Hanieh, *Carbon Capitalism*; Malm and Zetkin Collective, *White Skin, Black Fuel.*

85 Huber, *Lifeblood*, 98–127.

86 Chatterjee, 'Towards an Energetics of Class', 545.

87 Pirani, *Burning Up*, 99 (emphasis added).

88 Jeremy Walker, *More Heat than Life: The Tangled Roots of Ecology, Energy, and Economics* (Springer Nature, 2020), 12.

89 Phillipe Roqueplo, 'Der saure Regen: Ein "Unfall in Zeitlupe". Ein Beitrag zu einer Soziologie des Risikos', *Soziale Welt*, 37.4 (1986), 402–26.

90 Mattioli et al., 'The Political Economy of Car Dependence'.

91 Hobsbawm, *The Age of Extremes*, 277, 280.

92 Mattioli et al., 'The Political Economy of Car Dependence'; Pirani, *Burning Up*.

93 Pirani, *Burning Up*, 44.

94 Ståle Holgersen, 'What Are We Waiting For?', *Spectre*, 2025.

95 Pirani, *Burning Up*.

96 Marshal Berman, *All That Is Solid Melts into Air: The Experience of Modernity* (Verso, 2010), 293, 295.

97 Edgerton, *The Rise and Fall of the British Nation*, 319.

98 Paterson, *Automobile Politics*.

99 Quoted in Edensor, *National Identity*, 124.

100 Jim Tomlinson, *The Politics of Decline: Understanding Postwar Britain* (Longman, 2001).

101 Deindustrialisation is used here to denote how industrial employment came to represent a smaller share of total employment in many countries. Jim Tomlinson, 'Re-inventing the "Moral Economy" in Post-war Britain', *Historical Research*, 84 (2011); Erik Baker, *Make Your Own Job: The Entrepreneurial Work Ethic in Modern America* (Harvard University Press, 2025).

102 Hall, 'The Great Moving Right Show' in *Selected Political Writings*, 79.

103 Nancy Fraser, *Cannibal Capitalism: How Our System Is Devouring Democracy, Care, and the Planet – and What We Can Do about It* (Verso, 2023).

104 Tom Haines-Doran, 'The Financialisation of Car Consumption', *New Political Economy*, 29.3 (2024).

105 Christopher Niedt and Brett Christophers, 'Value at Risk in the Suburbs: Eminent Domain and the Geographical Politics of the US Foreclosure Crisis', *International Journal of Urban and Regional Research*, 40.6 (2016).

106 Kaleb B. Nygaard, 'The Rescue of the US Auto Industry, Module B: Restructuring General Motors Through Bankruptcy', *Journal of Financial Crises*, 4.1 (2022).

107 Tom Haines-Doran, 'The Financialisation of Car Consumption', *New Political Economy*, 29.3 (2024).

108 Alex Callinicos, *The New Age of Catastrophe* (Polity Press, 2023), 120.

109 Liam Stanley, *Britain Alone: How a Decade of Conflict Remade the Nation* (Manchester University Press, 2022).

110 Tom Hazeldine, 'Revolt of the Rustbelt', *New Left Review*, 105 (2017), 54.

111 Karl Marx and Friedrich Engels, *The Communist Manifesto* (Pelican Books, 1972), 84.

112 For reflections on the symbolism of this image and quote, see Paterson, *Automobile Politics*, 1–4.

113 Gloria Li and Edward White, 'The Relentless Innovation Fuelling China's "Brutal" Car Wars', *Financial Times*, 6 April 2025.

114 Thierry Theurillat, 'Urban Growth, from Manufacturing to Consumption and Financialization: The Case of China's Contemporary Urban Development', *Regional Studies*, 56.8 (2022).

115 Ibid.

116 Harvey, *Rebel Cities*, 104.

117 The carbon cost of this endeavour needs to be tallied. China manufactures half the world's cement, required for concrete, depending on huge amounts of coal. Half of this cement is used for domestic transport infrastructure. See Fressoz, *More and More and More*.

118 Paolo Gerbaudo, 'The Electric Vehicle Developmental State', *Phenomenal World*, 11 April 2024.

119 Tim Sahay, 'The Belt and Road 2.0: An Interview with Mathias Larsen on China's Overseas Clean-Tech Manufacturing Investments', *Phenomenal World*, 19 September 2025.

120 Ilaria Mazzocco and Ryan Featherston. 'The Global EV Shift: The Role of China and Industrial Policy in Emerging Economies', *Center for Strategic & International Studies*, 2025.

121 Gerbaudo, 'The Electric Vehicle Developmental State'.

122 Jasper Jolly, 'Sing When You're Winning: How Karaoke in Cars Heralds the Triumph of Chinese Firms', *Guardian*, 20 May 2025.

123 Sander Tordoir and Brad Setser, *How German Industry Can Survive the Second China Shock* (Centre for European Reform, 2025), 1.

124 Benjamin Wray, 'China and the Geopolitics of the Green Transition', *Transnational Institute*, 4 February 2025.

125 Wolfgang Münchau, *Kaput: The End of the German Miracle* (Swift Press, 2024), 52.

126 Alberto Toscano, *Late Fascism* (Verso, 2024).

3. Lockdown City

1 Brown, *In the Ruins*, 177.

2 Stuart Hall, 'The Great Moving Right Show', *Marxism Today*, 23.1 (1979), 19.

3 Nathalie Ortar and Patrick Rérat, eds, *Cycling Through the Pandemic: Tactical Urbanism and the Implementation of Pop-Up Bike Lanes in the Time of COVID-19* (Springer Nature, 2024), 4.

4 Sebastian Kraus and Nicolas Koch, 'Provisional COVID-19 Infrastructure Induces Large, Rapid Increases in Cycling', *Proceedings of the National Academy of Sciences*, 118.15 (2021).

5 Kate Belcher, Ceri Davies, Ella Guscott, Eliska Holland and Joshua Vey, *Low Traffic Neighbourhoods Research Report* (European Climate Foundation, 2021).

6 For more see Rebekah Diski, 'The Full Force of the Law: Protest and Repression on a Scorched Earth', *Salvage*, 15 (2025)
7 The Canary, 'Court of Appeal Erodes Independence of Juries', *Canary*, 20 January 2026.
8 M. E. O'Brien, *Family Abolition: Capitalism and the Communizing of Care* (Pluto Press, 2023), 26.
9 See Alva Gotby, 'The Labour of Lesbian Life: Wages Due Lesbians and the Politics of Refusal', *Salvage*, 13 (2024), 79.
10 O'Brien, *Family Abolition*, 153.
11 Ibid., 158.
12 Huber, *Lifeblood*, 82.
13 Sophie Lewis, *Abolish the Family: A Manifesto for Care and Liberation* (Verso, 2022).
14 Cara Daggett, 'Petro-masculinity: Fossil Fuels and Authoritarian Desire', *Millennium*, 47.1 (2018), 27–9.
15 Tim Burrows, *The Invention of Essex: The Making of an English County* (Profile Books, 2023).
16 Simon Heffer, 'Essex Man', *Sunday Telegraph*, 7 October 1990.
17 Danny Dorling, 'Dying Quietly: English Suburbs and the Stiff Upper Lip', *Political Quarterly*, 90.1 (2019).
18 Giulio Mattioli et al. 'Transport Poverty and Fuel Poverty in the UK: From Analogy to Comparison', *Transport Policy*, 59 (2017).
19 Dorling, 'Dying Quietly', 32.
20 Ben Ansell and David Adler, 'Brexit and the Politics of Housing in Britain', *Political Quarterly*, 90.S2, (2019).
21 Dorling, 'Dying Quietly', 37.
22 See Daniel Denvir, 'Counterrevolution w/ Melinda Cooper', *The Dig Podcast*, 30 September 2025; David Adler and Ben Ansell, 'Housing and Populism', *West European Politics*, 43.2 (2020).
23 Noam Gidron and Peter A. Hall, 'The Politics of Social Status: Economic and Cultural Roots of the Populist Right', *British Journal of Sociology*, 68.S1 (2017).
24 Benedicta Marzinotto, 'Trade Shocks and Relative Consumption: Why the European Middle Class Is Turning Far Right', *Review of International Political Economy* (2025).
25 Richard Seymour, *Disaster Nationalism: The Downfall of Liberal Civilization* (Verso, 2024), 40.
26 Adler and Ansell, *Housing and Populism*.
27 See Wendy Brown, *In the Ruins of Neoliberalism: The Rise of Antidemocratic*

Politics in the West (Columbia University Press, 2019); Michel Feher, *Producteurs et parasites. L'imaginaire si désirable du Rassemblement national* (La Découverte, 2024); Biko Koenig and Tali Mendelberg, 'The Symbolic Politics of Status in the MAGA Movement', *Perspectives on Politics* (2025).

28 Phil A. Neel, 'The New Geography of Suburbia: An Anatomy of America's Hinterland', *New Labor Forum*, 14 May 2018.

29 Ellie O'Donnell, Richa Syal and Alice Ross, 'Conservative-Run Anti-Ulez Facebook Groups Hosted Racist and Islamophobic Posts', *Unearthed*, 27 April 2024, at unearthed.greenpeace.org.

30 Johnson had driven a JCB through a mock wall bearing the slogan 'Get Brexit Done'. For a discussion of the scene, see Malm and Zetkin, *White Skin, Black Fuel*, 78.

31 Stuart Hall, 'The Great Moving Right Show', 19.

32 Jessica Murray and Robert Booth, 'UK a "Powder Keg" of Social Tensions a Year on from Summer Riots, Report Warns', *Guardian*, 15 July 2025.

33 Sanya Burgess, 'British Vigilantes Slash Small Migrant Boats on French Coastline', *iNews*, 19 November 2025.

4. Metabolism of the Nation

1 Jacob McLean, 'Carbon Convoys: Extractive Populism and the Canadian Far Right', doctoral dissertation, York University, Toronto, 2025. See also Jordan Kinder, *Petroturfing: Refining Canadian Oil through Social Media* (University of Minnesota Press, 2024); Tanner Mirrlees, 'The Carbon Convoy: The Climate Emergency Fueling the Far Right's Big Rigs', *Energy Humanities*, 3 May 2022.

2 Elizabeth Chatterjee, 'Towards an Energetics of Class: Comparing Energy Protests in India and the United States', *Comparative Studies in Society and History*, 66.3 (2024).

3 Marco D'Eramo, 'L'Europe profonde', *NLR/Sidecar*, 14 March 2024.

4 'Disorderly emissions' is Ernst Bloch's memorable coinage describing the 'non-synchronous' ideological elements on which far right ideology draws, 'fraudulently reactivat[ing] unfulfilled pasts and unrealised futures'. See Alberto Toscano, *Late Fascism: Race, Capitalism and the Politics of Crisis* (Verso, 2023), 7; Ernst Bloch and Mark Ritter, 'Nonsynchronism and the Obligation to Its Dialectics', *New German Critique*, 11 (1977), 38.

5 Arjen Siegmann, 'The Farmers' Revolt in the Netherlands: Causes and Consequences', *European View*, 23.2 (2024).

6 D'Eramo, 'L'Europe profonde'.

7 Chatterjee, 'Towards an Energetics of Class', 537.

8 Simon Enoch, 'The Oil Industry's Frankenstein', *Briarpatch*, 14 October 2022.

9 McLean, 'Carbon Convoys'.

10 Canadian Anti-Hate Network, 'Freedom Convoy Aligning with Farmer Protests in the Netherlands', *Canadian Anti-Hate Network*, 26 July 2022.

11 Bàrbara Molas, 'Dutch Flags and Maple Leaves: How Conspiracy Theories Created a Transnational Far-Right', *International Centre for Counter-Terrorism* (ICCT), 10 October 2022.

12 D'Eramo, 'L'Europe profonde'.

13 Roos Saat, 'The Making of Regressive Rural Populism in the Netherlands', unpublished MA thesis, International Institute for Social Studies, 2023, 20–1.

14 Matthew Levinger and Paula Franklin Lytle, 'Myth and Mobilisation: The Triadic Structure of Nationalist Rhetoric', *Nations and Nationalism*, 7.2 (2001).

15 Stuart Hall, 'On Postmodernism and Articulation: An Interview with Stuart Hall by Larry Grossberg and Others', in *Essential Essays; Volume 1: Foundations of Cultural Studies*, ed. by David Morley (Duke University Press, 2019), 238.

16 George Monbiot, 'Agrarian Populism', *George Monbiot*, 24 January 2024, at monbiot.com. On 'petro-farming', see Adam Hanieh, *Crude Capitalism: Oil, Corporate Power, and the Making of the World Market* (Verso, 2024).

17 Eimear Mc Loughlin, 'Protesting the Future: The Evolution of the European Farmer', *Anthropology Today*, 40.5 (2024); Jan Douwe van der Ploeg, 'Farmers' Upheaval, Climate Crisis and Populism', *Journal of Peasant Studies*, 47.3 (2020), 6.

18 D'Eramo, 'L'Europe profonde'.

19 Bloch and Ritter, 'Nonsynchronism and the Obligation to Its Dialectics', 26.

20 Giovanni Zibordi, 'Towards a Definition of Fascism', in David Beetham, ed., *Marxists in Face of Fascism: Writings by Marxists on Fascism from the Inter-war Period* (Manchester University Press, 1983).

21 Roger Cohen and Ivor Prickett, 'Angry Farmers Are Reshaping Europe', *New York Times*, 31 March 2024.

22 For a full biography of Morana's climate stance, see his extensive file at *DeSmog*.

23 Rachel Sherrington, Katharina Wecker, Laura Villadiego, Kevin Carboni, Marta Kasztelan, Coen Ramaer and Clare Carlile, 'The EU Elections Candidates Spreading False Narratives on Food and Farming', *DeSmog*, 6 June 2024.

24 Jonathan Haidt, Paul Rozin, Clark McCauley and Sumio Imada, 'Body, Psyche, and Culture: The Relationship between Disgust and Morality', *Psychology and Developing Societies*, 9.1 (1997).

25 Erin E. Buckels and Paul D. Trapnell, 'Disgust Facilitates Outgroup

Dehumanization', *Group Processes and Intergroup Relations*, 16.6 (2013). Susan B. Miller *Disgust: The Gatekeeper Emotion* (Taylor and Francis, 2013), 154.

26 Buckels and Trapnell, 'Disgust Facilitates Outgroup Dehumanization'.

27 Jonathan Davies, 'Brexit and Invasive Species: A Case Study of the Cognitive and Affective Encoding of "Abject Nature" in Contemporary Nationalist Ideology', *Cultural Studies*, 36.4 (2021), 569, 571, 576.

28 Sivamohan Valluvan, *Clamour of Nationalism: Race and Nation in Twenty-First-Century Britain* (Manchester University Press, 2019), 131.

29 The Lotus Eaters are inhabitants of a remote island in Homer's *Odyssey*, surviving on insects which also produce a pacifying effect that nullifies a sense of purpose, be it civic duty or personal commitments.

30 Jessica Murray, 'Nazi-Obsessed Terrorist Given Life Sentence for Worcestershire Attack on Asylum Seeker', *Guardian*, 17 January 2025.

31 The Farmer's Defence Force was a key coordinator of many of the roadblocks in the Netherlands. The organisation grew out of quasi-vigilante efforts to physically defend farms from militant animal rights groups and was the recipient of a five-figure sum from one agribusiness company. See Van der Ploeg, 'Farmers' Upheaval, Climate Crisis and Populism', 598.

32 Michael Billig, *Banal Nationalism* (Sage, 1995); A. Ichijo, 'Food and Nationalism: Gastronationalism Revisited', *Nationalities Papers*, 48.2 (2020), 215–23.

33 Harriet Bergman, 'Unite the White', *Medium*, 12 January 2024, at medium.com.

34 In *The Road to Wigan Pier* George Orwell notes how socialism and communism attract 'every fruit-juice drinker, nudist, sandal-wearer, sex-maniac, Quaker, "Nature Cure" quack, pacifist, and feminist in England'.

35 Chris Otter, 'Milk in Motion: Logistical Geographies in Twentieth-Century Britain', *Global Food History*, 9.1 (2023).

36 Derek Beres, Matthew Remski and Julian Walker, *Conspirituality: How New Age Conspiracy Theories Became a Health Threat* (Random House, 2023); MAHA is Robert F. Kennedy Jr's campaign to 'Make America Healthy Again'.

37 S. Marek Muller, David Rooney and Cecilia Cerja, 'Long Live the Liver King: Right-Wing Carnivorism and the Digital Dissemination of Primal Rhetoric', *Frontiers in Communication*, 9 (2024).

38 The latter was revealed by Hope Not Hate to be Charles Cornish-Dale, a thirty-something Briton with a PhD from Oxford. BAP's real name is Costin Vlad Alamariu, a Romanian-American with a PhD from Yale in political science.

39 Joshua Molloy and Eviane Leidig, 'The Emerging Raw Food Movement and the "Great Reset"', *Global Network on Extremism and Technology*, 10 October 2022, at gnet-research.org.

40 Josh Vandiver, 'Metapolitics, Masculinity, and Technology in the Rise of "Bronze Age Pervert"', in A. James McAdams and Alejandro Castrillon, eds, *Contemporary Far-Right Thinkers and the Future of Liberal Democracy* (Routledge, 2021).

41 Bronze Age Pervert, *Bronze Age Mindset: An Exhortation* (Self-published, 2018), Chapter 27.

42 Klaus Theweleit, *Male Fantasies: Women, Floods, Bodies, History*, vol. 1 (University of Minnesota Press, 1987).

43 Van der Ploeg, 'Farmers' Upheaval, Climate Crisis and Populism', 596.

44 Huber, *Lifeblood*, 88.

45 For more on ecofascism as a de facto form of climate denial 'by stealth', see Lise Benoist, Joe Turner, and Daniel Bailey, 'Ecofascism in the Shadow of "Patriotic Ecology": Nativism, Economic Greenwashing, and the Evolution of Far-Right Political Ecology in France', *Politics*, 46.1 (2024).

5. American Auto Apocalypse

1 Seth Schindler et al., 'The Second Cold War: US–China Competition for Centrality in Infrastructure, Digital, Production, and Finance Networks', *Geopolitics*, 29.4 (2023); Zeyi Yang, 'How Did China Come to Dominate the World of Electric Cars?', *MIT Technology Review*, 21 February 2023.

2 Largely due to the shale oil boom initiated under Obama, the quantity of crude oil exports has increased dramatically over the last decade, with new records typically broken each subsequent year.

3 Alexander Hertel-Fernandez, Alex Kiefel and Alan Yan, *The Varied Voice of Labor: Unpacking the Political Engagement of Labor in the 2024 Election* (Center for Labor and a Just Economy, 2025); Justin Mayhugh, 'UAW Poll Shows Member Support for Harris Growing', *United Auto Workers*, 23 October 2024, at uaw.org.

4 Alberto Toscano, *Late Fascism: Race, Capitalism and the Politics of Crisis* (Verso, 2023), 2.

5 Aaron Pelo, 'Trump's "Support" for Workers a Fake Enterprise', *MI AFL-CIO*, 5 February 2024, at miaflcio.org.

6 Leo Löwenthal and Norbert Guterman, *Prophets of Deceit: A Study of the Techniques of the American Agitator* (Verso, 2021 [1949]), 158.

7 Ibid., 128.

8 Ibid., 66.

9 Ibid., 128.

10 Hall et al., *Policing the Crisis*, 222–4.

11 Ibid., 323.
12 Wendy Brown, *In the Ruins of Neoliberalism: The Rise of Antidemocratic Politics in the West* (Columbia University Press, 2019), 164.
13 Julie Livingston and Andrew Ross, *Cars and Jails: Freedom Dreams, Debt, and Carcerality* (OR Books, 2022), 43.
14 Eric Levenson, 'Car Crashes and Rammings Take Center Stage in Immigration Crackdown', *CNN*, 25 October 2025.
15 Ernst Bloch quoted in Toscano, *Late Fascism*, 7.
16 Löwenthal and Guterman, *Prophets of Deceit*, 127.
17 Coral Davenport, 'G.M. Sold Millions of Cars That Were More Polluting than Allowed, E.P.A. Says', *New York Times*, 3 July 2024.
18 Coral Davenport, 'US to Announce Rollback of Auto Pollution Rules, a Key Effort to Fight Climate Change', *New York Times*, 30 March 2020; Trevor Houser, Kate Larsen, John Larsen, Peter Marsters and Hannah Pitt, *The Biggest Climate Rollback Yet?* (Rhodium Group, 2018).
19 Coral Davenport, 'Inside Trump's Plan to Halt Hundreds of Regulations', *New York Times*, 15 April 2025.
20 Coral Davenport, 'Biden Administration Announces Rule Aimed at Expanding Electric Vehicles', *New York Times*, 20 March 2024.
21 John Bozzella, 'EPA's Final EV Plan: The Ragged Edge of Achievable?', *Alliance for Automotive Innovation*, 20 March 2024, at autosinnovate.org.
22 'Auto Tariff Executive Order Brings Relief: John Bozzella', *Balance of Power*, Bloomberg TV, 29 April 2025.
23 Doug Henwood, 'Bidenomics Puts Business, Not Workers, First', *Jacobin*, 24 December 2023.
24 Patrick Bigger, Johanna Bozuwa, Mijin Cha, Daniel Aldana Cohen, Billy Fleming, Yonah Freemark, Batul Hassan, Mark Paul and Thea Riofrancos, *Inflation Reduction Act: the Good, the Bad, the Ugly* (Climate and Community Institute, 2022).
25 Brett Christophers, 'Failure as Success: What is Biden's Climate Legacy?', *Breakdown*, 1 (2025), 25.
26 Kana Inagaki, Harry Dempsey, Edward White and Sebastien Ash, 'Why Carmakers Are Falling Back in Love with Petrol', *Financial Times*, 8 October 2025.
27 Carroll, *Regime of Obstruction*.
28 Kalea Hall and David Shepardson, 'GM to Invest $4B in 3 U.S. Facilities over Next 2 Years as It Ramps Up Gas-Powered Vehicles', *CBC News*, 11 June 2025; Sharon Terlep, Becky Peterson and Lindsay Wise, 'GM Is Pushing Hard to Tank California's EV Mandate', *Wall Street Journal*, 18 May 2025.

29 Inagaki et al., 'Why Carmakers Are Falling Back in Love with Petrol'.

30 Malm and Zetkin, *White Skin, Black Fuel*, 185.

31 Chris Kirkham, 'Musk Embraces Trump and Scorns Subsidies. But Tesla Still Lobbies for US Benefits', *Reuters*, 12 August 2024.

32 Denialism could reign supreme in the US in part because 'panels and turbines had no Koch brothers'. See Malm and Zetkin, *White Skin, Black Fuel*, 205.

33 See Brett Christophers, *The Price Is Wrong: Why Capitalism Won't Save the Planet* (Verso, 2024).

34 Coralie Kraft, 'Secret Tunnels, Bunkers and Arsenals: The "Panic Industry" Is Booming', *New York Times*, 10 April 2025.

35 Ulrich Brand and Markus Wissen, *The Imperial Mode of Living: Everyday Life and the Ecological Crisis of Capitalism* (Verso, 2021).

36 George Hammond and Stephen Morris, 'Elon Musk to Seek Tesla Board Approval for $5bn Injection into xAI Start-Up', *Financial Times*, 25 July 2024.

37 While the sale of regulatory credits accounts for a rather small portion of the firm's revenues, these sales have no cost attached; that is, credits are free to produce since Tesla receives them automatically from selling vehicles that exceed emissions standards. In other words, they are 'pure profit'. Therefore, while only accounting for about 3 per cent of Tesla's revenue in the decade between 2014 and 2024, sale of these credits, amounting to $10.7 billion, accounted for 32 per cent of the company's profits over the same period. In the first quarter of 2024, the company would have made a loss if not for these credits. See Stephen Morris and Alex Rogers, '"Elon Has Finally Woken Up": Musk Battles to Save Tesla from Trump', *Financial Times*, 4 July 2025.

38 Marc Andreessen, 'The Techno-optimist Manifesto', *Andreessen Horowitz*, 16 October 2023, at a16z.com.

39 Malm and Zetkin Collective, *White Skin, Black Fuel*, 400–9.

40 Steve Bannon: 'President Trump Will Serve a Third Term', dir. by *Financial Times* (2025).

41 Kat Lonsdorf, 'What Trump's national emergencies could mean for American democracy', *NPR*, 9 June 2025.

42 Melinda Cooper, 'Trump's Antisocial State', *Dissent Magazine*, 18 March 2025.

43 Ibid.

44 One public tracker indicates that nearly half of Project 2025's policy recommendations have been implemented in under a year. See project2025.observer/en.

45 Remarkably, a fraction of Democrats (35 of 212) joined them in this, framing EVs as too expensive and impractical, suggesting that the years-long campaign against EVs was starting to shift liberal common sense.

46 Kate Aronoff, 'Republicans Are Rewriting Congressional Rules so Cars Can Pollute More', *New Republic*, 22 May 2025.

47 Olivia Guarna and Michael Burger, 'Demystifying President Trump's "National Energy Emergency" and the Scope of Emergency Authority', *Climate Law Blog*, Sabin Center for Climate Change Law, Columbia Law School, 14 February 2025.

48 Ben King, John Larsen, Nathan Pastorek, Hannah Kolus, Anna van Brummen and Michael Gaffney, *Three Key Outcomes of the 'One Big Beautiful Bill Act' on US Manufacturing and Innovation* (Rhodium Group, 2025).

49 Hall et al., *Policing the Crisis*, 323.

50 Ibid., 217.

51 Ibid., 217.

52 Nicos Poulantzas, *Fascism and Dictatorship: The Third International and the Problem of Fascism* (Verso, 2019), 298.

53 Geoff Mann and Joel Wainwright, 'Political Scenarios for Climate Disaster', *Dissent Magazine*, Summer (2019).

54 See Aziz Rana, *The Constitutional Bind: How Americans Came to Idolize a Document That Fails Them* (University of Chicago Press, 2024).

55 Robert O. Paxton, *The Anatomy of Fascism* (Vintage Books, 2005), 218–19.

6. The Remigration Industrial Complex

1 Gaia Vince, *Nomad Century* (Penguin Books, 2023), xiv.

2 Chi Xu, Timothy Kohler, Timothy Lenton, Jens-Christian Svenning and Martin Scheffer, 'Future of the Human Climate Niche', *Proceedings of the National Academy of Sciences*, 17.21 (2020).

3 Malm and Zetkin, *White Skin, Black Fuel*; see the chapter 'Ecology Is the Border'.

4 Robert Kiesel and Maria Fiedler, 'Nach der Europawahl: AfD-Jugend entdeckt das Klima', *Tagesspiegel*, 28 May 2019.

5 Tatjana Söding and William Callison, 'Postwachstum von rechts und die Gefahr des Ökofaschismus', *Ökologisches Wirtschaften*, 38.1 (2023).

6 Junge Alternative, *Jugend, die vorangeht! Programm und Leitlinien* (Junge Alternative für Deutschland, 2023).

7 Janet Biehl and Peter Staudenmaier, *Ecofascism Revisited: Lessons from the German Experience* (New Compass Press, 2011).

8 Bernhard Forchtner, *The Far Right and the Environment: Politics, Discourse and Communication* (Routledge, 2019).

9 Katy Fallon and Giorgos Christides, '"Fire on Fire": How Migrants Got Blamed for Greece's Devastating Blazes', *Guardian*, 23 November 2023.

10 European Data Journalism Network, 'Fires in Greece: When Migrants Become Scapegoats', 17 January 2024, at europeandatajournalism.eu.

11 Forensic Architecture, 'The Evros/Meriç River: A Century of Border Design', 15 January 2025.

12 Florent Marcellesi, 'Flooded with Lies: Climate Infodemic in Valencia', *Heinrich-Böll-Stiftung European Union*, 6 December 2024, at eu.boell.org.

13 Rachel Leingang, 'US Right Wing Fans Misinformation Fires as Firefighters Battle Los Angeles Blazes', *Guardian*, 10 January 2025.

14 Mimi Sheller, *Mobility Justice* (Verso, 2018), 1.

15 Eva von Redecker, *Bleibefreitheit* (S. Fischer, 2023).

16 Marcus Bensmann, Justus von Daniels, Anette Dowideit, Jean Peters and Gabriela Keller, 'Secret Plan against Germany', *Correctiv*, 15 January 2024.

17 Oliver Klein, Nils Metzger and Jan Henrich, 'AfD-Vertreter bekräftigen Pläne für "Remigration" – Programm', *ZDF heute*, 12 January 2024.

18 The AfD was not the first party to integrate demands for 'remigration' in their election programme. Following its identitarian coinage in France and Austria, for example, Wilders's PVV proposed to establish a 'Ministry of Immigration, Remigration and De-Islamization'.

19 In this instance of 'weaponised automobility', the improper use of the nation's technology by foreigners is a national emergency. See Satya Savitzky and Julie Cidell, 'Whose Streets? Roadway Protests and Weaponised Automobility', *Antipode*, 55.5 (2023).

20 Ashifa Kassam, 'How "Remigration" Became a Buzzword for Global Far-Right', *Guardian*, 3 October 2024. For a study of far-right mainstreaming, discussed in this section, see Katy Brown, Aurelien Mondon, Aaron Winter, 'The far right, the mainstream and mainstreaming: towards a heuristic framework', *Journal of Political Ideologies*, 28.2 (2023).

21 For more on femonationalism see S. R. Farris, *In the Name of Women's Rights: The Rise of Femonationalism* (Duke University Press, 2017).

22 Gisela Mackenroth, 'Rettet den Diesel!', *Gesellschaft unter Spannung*, 40.1 (2021).

23 Oleksiy Radynski, 'Russian Fossil Fascism Is Europe's Fault', *Soniakh*, 4 October 2022.

24 Max Hägler, 'Auto Shanghai 2023: Fahrzeuge chinesischer Firmen, überall', *Die Zeit*, 20 April 2023.

25 Wolfgang Münchau, *Kaput* (Swift Press, 2024).

26 James Rothwell, 'German Opposition Leaders Vow to fight EU Car-Ban Plans', *Daily Telegraph*, 12 March 2024.

27 Matthew Karnitschnig, 'Rust Belt on the Rhine', *Politico*, 13 July 2023.

28 Münchau, *Kaput*.

29 Arató László, 'Germany's Auto Industry Prepares for Job Losses in Electric Transition', *Euronews*, 30 October 2024.

30 Doloresz Katanich, 'Volkswagen Strike: "Still a Long Way from a Viable Solution"', *Euronews*, 10 December 2024.

31 Michelle Pace, 'German Election: Far-Right Firewall Weakens as Immigration Concerns Take Centre Stage', *Chatham House – The Royal Institute of International Affairs*, 20 February 2025, at chathamhouse.org.

32 Siri Aas Rustad, 'Conflict Trends: A Global Overview, 1946–2024', *PRIO*, 2025, at prio.org.

33 Stuart Parkinson and Linsey Cottrell, *Estimating the Military's Global Greenhouse Gas Emissions* (Scientists for Global Responsibility and Conflict and Environment Observatory, 2022).

34 Andreas Malm, *The Destruction of Palestine Is the Destruction of the Earth* (Verso, 2025).

35 Benjamin Neimark, Frederick Otu-Larbi, Frederick, Reuben Larbi Patrick Bigger, Linsey Cottrell, Lennard de Klerk and Mykola Shlapak, 'War on the Climate: A Multitemporal Study of Greenhouse Gas Emissions of the Israel-Gaza Conflict', *Social Science Research Network* (2025), at papers.ssrn.com.

36 Lennard de Klerk, Mykola Shlapak, Sergiy Zibtsev, Viktor Myroniuk, Oleksandr Soshenskyi, Roman Vasylyshyn, Svitlana Krakovska and Lidiia Kryshtop, *Climate Damage Caused by Russia's War in Ukraine, 24 February 2022–23 February 2025: Preliminary Assessment* (Initiative on GHG Accounting of War, 2025).

37 Nina Lakhani 'Emissions from Israel's War in Gaza Have "Immense" Effect on Climate Catastrophe', *Guardian*, 9 January 2024.

38 European Commission and High Representative, 'White Paper for European Defence – Readiness 2030', *Defence Industry and Space*, 12 March 2025.

39 'Union und SPD einigen sich auf Milliardenkredite für Verteidigung und Infrastruktur: Sondierungsergebnis Finanzen', *Tagesschau*, 8 March 2025.

40 Xiao Liang, Nan Tian, Diego Lopes da Silva, Lorenzo Scarazzato, Zubaida Karim and Jade Guiberteau Ricard, *Trends in World Military Expenditure, 2024*, SIPRI fact sheet, 2025.

41 Lulu Garcia Navarro, 'The Interview: Mark Rutte, the Head of Nato Thinks President Trump Deserves Praise', *New York Times Magazine*, 5 July 2025.

42 Matt Phillips, 'Why Trump's Election Win Is Sending Shares of a German Arms Firm Soaring: German Weapons Giant Rheinmetall Jumps after U.S. Election Shock', *Sherwood News*, 8 November 2024.

43 EY and Dekabank, 'Wirtschaftliche Effekte europäischer Verteidigungsinvestitionen', 12 February 2025.

44 'Rheinmetall hält Rückkehr zur Wehrpflicht für wahrscheinlich', *Regional Heute*, 14 March 2025.

45 CDU/CSU and SPD, *Koalitionsvertrag 'Verantwortung für Deutschland': Koalitionsvereinbarung für die 21. Legislaturperiode* (2025).

46 'Germany's Shrinking Auto Industry May Be Key to Defence Ramp Up', *Deutsche Bank Research*, 31 March 2025, at dbresearch.com.

47 Ethan Ilzetzki, *Guns and Growth: The Economic Consequences of Defense Buildups* (Kiel Institute for the World Economy, Kiel Report No. 2, 2025).

48 Ryan Bourne, 'No, Sir Keir Starmer, Defence Spending Won't Make Britain Richer', *Cato Institute – Commentary*, 12 March 2025, at cato.org.

49 Joise Ensor, 'Cuts to USAID Blamed for 300,000 Deaths – Most of Them Children', *The Times*, 30 May 2025.

50 Daniella Medeiros Cavalcanti, Lucas de Oliveira Ferreira de Sales, Andrea Ferreira da Silva, Elisa Landin Basterra, Daiana Pena and Caterina Monti, 'Evaluating the Impact of Two Decades of USAID Interventions and Projecting the Effects of Defunding on Mortality up to 2030: A Retrospective Impact Evaluation and Forecasting Analysis', *The Lancet*, 406 (2025).

51 US Department of State, *Making Foreign Aid Great Again* (2025).

52 Julia Frankel and Leo Correro, 'Far-Right Group Holds Antisemitism Conference in Israel', *AP News*, 12 November 2025.

53 Kate Brady and Aaron Wiener, 'German Far-Right Activist Seeks Asylum in U.S. as Trump Ties Deepen', *Washington Post*, 9 November 2025.

54 Ernst Fraekel, *The Dual State: A Contribution to the Theory of Dictatorship* (Oxford University Press, 2017).

55 Steffen Mau, *Sorting Machines: The Reinvention of the Border in the 21st Century* (Polity, 2022).

56 Garett Hardin, 'Lifeboat Ethics: the Case Against Helping the Poor', *Psychology Today*, 8 (1974), 800.

57 Zygmunt Baumann, *Globalisation: The Human Consequences* (Columbia University Press, 1988).